An Illustrated History
of Mayer, Arizona

ALSO BY NANCY BURGESS

*A Photographic Tour of 1916
Prescott, Arizona* (McFarland, 2005)

An Illustrated History of Mayer, Arizona

Stagecoaches, Mining, Ranching and the Railroad

NANCY BURGESS

Foreword by Marshall Trimble

McFarland & Company, Inc., Publishers
Jefferson, North Carolina, and London

LIBRARY OF CONGRESS CATALOGUING-IN-PUBLICATION DATA

Burgess, Nancy.
An illustrated history of Mayer, Arizona :
stagecoaches, mining, ranching and the railroad /
Nancy Burgess ; foreword by Marshall Trimble.
p.   cm.

Includes bibliographical references and index.

ISBN 978-0-7864-6287-2
softcover : acid free paper ∞

1. Mayer (Ariz.)—History.   2. Mayer (Ariz.)—History—Pictorial works.
3. Mayer (Ariz.)—Biography.   4. Mayer (Ariz.)—Economic conditions.
5. Stagecoach lines—Arizona—Mayer—History.   6. Mines and mineral resources—
Arizona—Mayer—History. 7. Ranching—Arizona—Mayer—History.
8. Railroads—Arizona—Mayer—History.   I. Title.
F819.M28B87 2012   979.1'57—dc23   2012014484

BRITISH LIBRARY CATALOGUING DATA ARE AVAILABLE

© 2012 Nancy Burgess. All rights reserved

*No part of this book may be reproduced or transmitted in any form
or by any means, electronic or mechanical, including photocopying
or recording, or by any information storage and retrieval system,
without permission in writing from the publisher.*

On the cover: This view of Mayer is taken facing southwest about 1900–1901,
as a Prescott & Eastern locomotive, tender, combination car and four passenger cars had
just crossed the trestle on the way to the depot (courtesy Arizona Historical Society, Tucson)

Manufactured in the United States of America

*McFarland & Company, Inc., Publishers
Box 611, Jefferson, North Carolina 28640
www.mcfarlandpub.com*

Contents

Acknowledgments	vii
Foreword by Marshall Trimble, Arizona State Historian	1
Preface	5
Introduction	7
1. The Black Canyon Road, or, How Do you Get There From Here?	9
2. The Mayer Family at Big Bug Station	25
3. The Town of Mayer: How the West Was Really Won	39
4. Ranchers and Farmers Follow the Dusty Trail to Mayer	54
5. School Bells Ring	75
6. Joe Mayer, Entrepreneur Extraordinaire	91
7. Business Is Booming	109
8. The Mining Boom: Boom and Bust	135
9. The Railroads Arrive	167
10. The Depression Years and World War II in Mayer	193
11. The Later Years: Post–World War II	215
12. The Modern Years: Mayer Today	238
Bibliography	255
Index	261

Acknowledgments

Writing about history is always a challenge. Sometimes there is too much information, sometimes not enough. Sometimes information and photographs are scattered among many repositories, including family archives. Sometimes one source has completely different information than another source. Sometimes the ever elusive detail that would solve a conflict just cannot be found. Sometimes the ever elusive photograph that the researcher knows is out there somewhere just cannot be found. Sometimes pure luck and serendipity are the solution to the mysteries that plague the historical researcher. All of these situations apply to *An Illustrated History of Mayer, Arizona*.

The initial collection of photographs, copies of photographs, newspapers, newspaper clippings, copies of newspaper clippings, letters and documents which form the core of this book came from the estate of Winifred (Winnie) Mayer Thorpe, Joe and Sarah Belle Mayer's youngest daughter. However, labeling and dating the items in this collection was not Winnie's strong suit.

Today, with the technology of the Internet, the task of researching historic events, places, people and stories is much easier and more efficient than in the past when "road trips" were a necessity. However, Internet information still has to be verified, cross-checked and corrected if necessary. Many archives are on the cutting edge of technology and have made incredible resources available to the searching public. Others are still doing it the old fashioned way—by hand, which occasionally results in a pleasant discovery. Online newspapers searchable by name or topic have saved many, many hours which would otherwise be spent looking at microfilm. But even with all of these resources, it takes people who are dedicated to collecting, preserving, transcribing, and cataloguing Arizona's precious historic resources to make it all work. In that regard, I would like to acknowledge with thanks the archivists and researchers who helped make these resources available from Sharlot Hall Museum, the Mayer Library, the Arizona Historical Foundation, the Arizona Mining and Mineral Museum in Phoenix, Hayden Library Special Collections at Arizona State University, and the Arizona Historical Society in Tucson.

I would also like to acknowledge and thank those individuals who went out of their way to help me in my search for the history of Mayer Arizona: my husband, Jim Burgess, who spent many, many hours "fixing" all of the images; Roselynn Fernwalt; Larry Howard; Kathy Lopez; Clyde and Sonja McDonald; Sylvia Neeley; Mike Connors; the Orr family and Bill Promberger; and, most of all, although she is no longer living, Winifred Mayer Thorpe.

All images which are not attributed in the caption are a part of the personal collection of the author. In some instances, others, including various archives, have the same photographs or images in their collections. The sources of all images, photographs, advertisements, business tokens, maps and copies from historic newspapers which have been provided by other published materials, institutional or private archives or collections are properly credited in the captions.

Foreword by Marshall Trimble

The earliest stagecoach road connecting Prescott with the old Gila Trail from Tucson to California headed south and west from Prescott, bypassing the tiny community of Phoenix. It had the advantage of flat desert country most of the way. At the time, Wickenburg was the hub of traffic to points west. But that changed when the future capital city began supplying large amounts of hay and food to Fort Verde and Prescott.

The idea for a wagon road from Phoenix to Prescott by way of Agua Fria River was first proposed in 1866. Both the army and local businessmen wanted a more direct road between the two towns.

In 1870 General George Stoneman, commanding officer of the new Arizona military district, determined to build a road from Fort McDowell north to Forts Whipple and Verde by way of Black Canyon. Near where Mayer is today the road forked, one leading west to Ft. Whipple and the other north to Ft. Verde.

At the same time, mines were opening throughout the Bradshaw Mountains.

Funds were raised, and in November 1873 the first wagon train completed the 95 mile journey on the new road in just five days, half the time it would have taken on the Wickenburg route.

In January 1878, a buckboard began carrying passengers north from Phoenix to Gillett. A one-way ticket cost five dollars and a round trip cost three dollars more. A month later, a stagecoach from Prescott made its first run headed for Gillett. A stagecoach could make the trip from Prescott to Phoenix in a day and a half; it took the heavy mule driven freight wagons a week, at 20 miles a day.

A one-way ticket on the stage ran a little higher, costing ten dollars. The Prescott-bound stage made stops at the New River station, the Agua Fria River crossing at Cañon (near today's Rock Springs), Bumble Bee, Cordes (Antelope), Mayer (Big Bug), Dewey (Spaulding's and Agua Fria), and Four-Mile Hill, just east of Prescott.

The old Black Canyon Stage Line was one of the most notorious in Arizona history for holdups and a favorite spot for highwaymen was on the steep grade east of Mayer leading into Crazy Basin. One of my favorites occurred in August 1882. Among the passengers were a Dr. Lloyd and a merchant named Isador Solomon. After checking the doctor's pockets and finding only thirty dollars the bandit remarked, "You are about as hard up as I am," and gave the money back. Solomon surrendered his gold watch and 55 dollars.

The robbers, who seemed to be relaxed old hands at robbing stages, talked and joked with the passengers as they were tearing open the strong box and Solomon took the opportunity to compliment them on their good behavior and then asked if he could have his watch back. The outlaw smiled and gave it back.

Another passenger asked if he could borrow a dollar for breakfast and instead was given two. Solomon then said he was also dead broke and was "loaned" three dollars.

About that time, the northbound stage from Phoenix approached and the highwaymen, armed with shotguns, brought it to a halt and ordered the passengers to get out.

After they'd gone through both strong boxes and cleaned out the passenger's pockets, the two bandits, wearing silk neckerchiefs to cover their faces, told everyone to get on board and get moving. As far as is known, the two old pros were never apprehended.

The Big Bug Station opened in 1879 and a year later William Muncey became owner of the station and 160 acres of land. Two years later, on July 13, 1882, he sold out to Joe Mayer for $1,200 in gold. Mayer's wife, Sadie, was carrying the gold from a caché stashed in the bottom of her sewing bag.

Mayer built a new station and home. The residence was divided into guest rooms at Joe and Sadie's residence. The station also sold general merchandise. There was also a small saloon where the passengers could imbibe, something that made the bumpy stage ride more bearable. Shady porches were built at the front and back. The station became well-known for its hospitality. When the post office opened in 1884, Sadie was the first postmistress. She also acted as town doctor.

Tragedy struck after placer miners dammed up Big Bug Creek six miles above Mayer Station. In 1891, the dam burst, sending a wall of water down the creek taking everything in its path, including the station. Mayer quickly rebuilt the station and soon he was putting up brick buildings. Mayer was becoming a bustling little town supplying the nearby mines. Joe assumed the role of "town father."

A major milestone for Mayer came in 1902 when Frank Murphy's "Impossible Bradshaw Mountain Railway" branched off from the Prescott & Eastern on its way up the mountain to Crown King. Steam-driven locomotives pulling trains carrying passengers, ore and mining equipment were chuffing through Mayer on a regular basis.

An enduring story in Mayer is that of an optimistic developer who planned to build a smelter in 1917 and contracted with a company to build a smelter and a smokestack. Unfortunately, the mine closed before the smelter went into operation. The developer tried to break the contracts. The smelter contractor acquiesced but the chimney company believed a deal was a deal and sent a crew out to complete the smokestack. Today a lonely non-smoking smokestack is perched on a hill overlooking the town.

One time a bunch of miners bet some cowboys that a woman working at the White House Hotel could climb the stack. Both sides put up $100. The miners won when she climbed to the top. This writer isn't sure if she got a share in the winnings but is willing to wager that she did.

During the late 1930s, plans were being made to build a highway along the old stage road from Phoenix to Prescott and Flagstaff. Plans were delayed with World War II, but in 1946 construction resumed. The old Black Canyon originally ran up Mission Drive, today's 27th Avenue, north through New River, Rock Springs, Bumble

Bee and Cordes, where it turned west to Mayer and Prescott. During the late 1950s, construction began to cut a road from the Agua Fria River crossing, up the mountain to Sunset Point, bypassing Bumble Bee, Cordes and Mayer. The road from Cordes Junction to Flagstaff wasn't completed until 1961.

Arizona songwriter Dean Cook once described to me how the family traveled from Flagstaff to Phoenix prior to the opening of the highway: "We went through Oak Creek and Jerome, to Prescott, then down through Mayer, Cordes and on to Phoenix." One had to have a real good reason to go to Phoenix in those days.

In 1947, my dad had just hired out as a fireman on the Santa Fe Railroad. We set out in a 1936 Ford, pulling a two-room trailer house up the Black Canyon Road on the way to our new home in Ash Fork. We opted for the Black Canyon route over the paved Wickenburg highway (knowing we'd never be able to pull that trailer up Yarnell Hill). The highway turned into a dirt road at New River Station. Our old car broke down on the steep grade south of Bumble Bee and had to be towed behind a Highway Department dump truck into Bumble Bee, where we took up residence until new parts could arrive from Phoenix. My brothers and I spent the days prospecting for gold in Big Bug Creek hoping that the parts would never arrive.

We set out again, and going up the treacherous hill from Crazy Basin to Cordes we broke down again. Henry Cordes was our host for a few days. It was smooth sailing on to Mayer and Prescott. I have vivid memories of driving through Mayer early one morning and being amazed at seeing the red brick buildings out in the middle of nowhere. I was eight years old and not in a hurry to get to Ash Fork (where I would have to enroll in school), and I remember hoping that old Ford would break down once again so we could live in Mayer for a few days.

It's great to see the histories of these small Arizona towns recorded for posterity as Nancy Burgess has in *An Illustrated History of Mayer, Arizona.* Mainstream history books generally focus on large communities, politicians and major events. The stories of these small towns is lost and gone forever if writers don't collect the photos and write the histories. Every community, no matter how small, has a history worth the telling.

Arizona native Marshall Trimble is the official state historian of Arizona. Often referred to as the "Will Rogers of Arizona," he grew up in the small town of Ash Fork. He has written more than twenty books on Arizona and the West, and is a well known television and radio host, storyteller, and teacher.

Arizona

DISTANCE FROM MAYER
Prescott 27 miles. Phoenix 75. Flagstaff 90. Tucson 190 miles

Preface

"The history of every country begins in the heart of a man or a woman."—Willa Cather, *Song of the Lark*

In her article entitled "Joe Mayer and His Town" written for the summer 1978 issue of the *Journal of Arizona History*, Winifred Mayer Thorpe described her father's arrival at Big Bug Stage Station:

> At the end of an exceptionally warm day in the spring of 1881,* a good-looking young rider with hazel eyes and black hair approached the Big Bug Stage Station on the banks of Big Bug Creek thirty-two miles southeast of Prescott. The place consisted only of a small shack, a corral and grove full of trees, but it was a welcome sight to Joe Mayer, who had cut across the mountains from the mining camp of Tip Top—a long, rough journey but shorter than the old Black Canyon stage route. His eyes rested on the surrounding mountains and the tiny stream of cool, clear water flowing between the thickly wooded banks. He told his family in later years that, at that moment, he knew he had come home.

The subject of this history is the small town of Mayer, Arizona. It is located near the center of the state at an elevation of 4,371 feet in the foothills of the Bradshaw Mountains. It was founded and named by Joseph Mayer and his wife, Sarah (Sadie) Belle Wilbur Mayer, who purchased the "Big Bug Stage Station," a stage stop on the Black Canyon Stage Line from Prescott to Phoenix, in 1882 for $1,200 in gold. Joe Mayer then proceeded to found a town, where he was the major player until his death in 1909.

Even after Joe Mayer's death, the Mayer family, which included daughters Mary Belle (Mamie), Martha Gertrude (Martie) and Winifred Lucille (Winnie), and son Wilbur Joseph (Burr), continued the family tradition of entrepreneurship and community service. Direct descendants of Joe and Sadie Mayer continued to live in Mayer until the death of Winifred's younger son in 2006. There are also other stories and characters to tell about in the history of Mayer, and many families who settled in Mayer decades ago are still there—an inextricable part of the history of a place and time in central Arizona which lives on today.

This book is not intended to be a comprehensive history of central Arizona. Nor is it intended to be a biography of the Mayer family, though the story of the man and

---

*Actually, 1882.

the town are inseparably interwoven. It is the story of a small, rural western town and the men and women who put it on the map. The early history of the area is told. The stories of the ranching, railroad and mining history are also told. The real story of Mayer, Arizona, is told partly through the words of Winifred Mayer Thorpe in her own voice, and through her collection of documents and photographs.

As the late Barry M. Goldwater wrote in 1961 in the publisher's note to *Arizona Territory Post Offices and Postmasters*, by John and Lillian Theobald, "historical research may be compared to winnowing grain as practiced by the primitive Indians of Arizona. After much stomping around on the grain, it was thrown into the air. Wind carried away some of the chaff. The nourishing kernels of grain fell back to the ground, to be tossed again and again into air until all the residue had blown away. Beneath the chaff, if one works hard enough, will be found the solid kernels of fact...." After much winnowing, it is hoped that this book presents as many solid kernels of fact about Joe and Sadie Mayer and Mayer, Arizona, as can be discovered today, along with a fascinating collection of photographs and documents.

Introduction

"History is more than a record of events—more than an accurate compilation of dates and names and places—true history is the most vivid picture of past conditions which we can project into the present and preserve for the future ... preserving to us the living men and women of the past and the scenes through which they moved, the difficulties they met and the materials out of which they built their tier in the great wall of life. No generation lives for itself alone, but tomorrow rests upon the shoulders of today—as today rests on the shoulders of yesterday."
—Sharlot M. Hall, from an untitled speech

The town of Mayer, Arizona, has had a varied history in Yavapai County but has retained its economic base, its rural, small-town, western country flavor and many of its historic buildings. As industries such as mining and sheep and cattle ranching, have waned, and as the highway has bypassed the downtown, and, as retirees and commuters from the big cities to the south have arrived, Mayer is still "Mayer," much as it was 50 years ago.

An Illustrated History of Mayer, Arizona is based on a collection of documents, photographs, maps and notes which belonged to Winifred Mayer Thorpe, Joe and Sarah Belle Wilbur (Sadie) Mayer's youngest daughter. This collection of materials was acquired at auction many years after Winifred's death. The author knew Mrs. Thorpe briefly before her death in July of 1983 and subsequently prepared nominations to the National Register of Historic Places for two buildings built by Joe Mayer and owned by Mrs. Thorpe—the Mayer Business Block and the Mayer Apartments.

There is no comprehensive written history of Mayer, and many of the images which illustrate the book have never been published. The history of the town of Mayer *can* be told with only the written word, however, the stories of Mayer's history are enhanced and expanded by the visual images in the form of photographs, maps, advertisements, business tokens and other printed material presented in *An Illustrated History of Mayer, Arizona*.

The present work is the result of more than twenty years of research. I am a historian and photographer specializing in the history of central Arizona and also the author of *A Photographic Tour of 1916 Prescott, Arizona* published by McFarland in 2005.

The history of Mayer covers broad topics which are integral to the history of central Arizona, and these topics are addressed to some extent in both text and visual images. These topics include ranching (cattle, angora goats and sheep), the railroad, and mining. Although the ranching tradition is still active in the Mayer area, it is today primarily focused on cattle. The railroad no longer runs to Mayer, but the importance of the railroad in the settlement and development of central Arizona, including many small towns such as Mayer, cannot be overemphasized. I have located the Mayer Depot, a key building in the economy and streetscape of Mayer in the days of the railroad, in Phoenix, Arizona. Mining is still a very viable industry in Arizona, in Yavapai County and in Mayer, although on a much smaller scale. All of these stories and more are presented in both text and historic and contemporary images.

Joe Mayer had his hand in almost every business in the Mayer area in the early days and was a true 19th century entrepreneur who was constantly coming up with new and adventuresome business ideas. Many of his enterprises were successful and made him a wealthy man in property, if not in cash. Some business ventures, such as his "cactus spine toothpicks" never took off. However, he helped to bring water and the railroad to Mayer, built the first hotel and the first business block, built the 1902 school and established the post office. Joe's wife, Sarah Belle (Sadie) Wilbur Mayer, was the first postmistress, starting in January of 1884, and she continued in that capacity into the 20th century. Sadie was also known throughout the area for her skill in "doctoring" and was always willing to help anyone who was in need of medical assistance in an area where the nearest doctor was some 25 or more miles away.

It is important to acknowledge and save the places that matter today and in our history. This book is the story of a small, rural western town and the men and women who put it on the map. There is common thread that runs through all of these people. They all have a true pioneering spirit, and the uncompromising land and lifestyle in which this type of person flourishes is called, even today, the American Frontier.

1

The Black Canyon Road, or, How Do You Get There from Here?

For thousands of years, the Wipukyipai (Yavapai) People lived in central Arizona in an area encompassing over ten million acres, including the area of the northeastern foothills of the Bradshaw Mountains in Yavapai County. Their homelands supported a population of over 1,000 people. A nomadic people who hunted and gathered and also grew some crops along the streams of central Arizona, they lived off the land and built temporary dwellings as they moved about the landscape. At one time there were four separate bands of the Yavapai, but they spoke the same Yuman-based language and considered themselves to be one people. Unfortunately, as Euro-Americans began to move into the traditional lands of the native people, they lost access to traditional hunting and gathering areas and to water. Conflicts ensued and quickly escalated. These conflicts culminated in the military campaign against the Indians led by General George S. Crook in the winter of 1872-73. Crook removed almost all Indians from Yavapai County by force-marching them to the San Carlos Indian Reservation in eastern Arizona in 1875. There, the Yavapai joined the Dilzhe'e (Tonto Apache) People, who were also native to the area. Different culturally from the Yavapai, the Apache speak an Athabascan-based language. Twenty years later, as the Yavapai and Apache people began to return to their homelands, they formed the nucleus of the three contemporary tribes. Today, the Fort McDowell Yavapai Nation, the "Abaja," is a 950-member tribe who live both on and off a 24,000-acre reservation in northeastern Maricopa County twenty-three miles northeast of Phoenix. The community was created by Executive Order on September 15, 1903. The Yavapai-Apache Nation is headquartered in Camp Verde. Recognized as a sovereign people by the United States government in 1932, the Nation is an amalgamation of the two ancient tribes: Wipukyipai (Yavapai) and Dilzhe'e (Tonto Apache). The Yavapai-Prescott Indian Tribe of less than 200 members lives near Prescott on a reservation of approximately 1,400 acres. The United States government recognized the Yavapai-Prescott Tribe's sovereignty on June 7, 1935. The members of these tribes take great pride in their history, culture and traditional values. The women of the tribes have a long tradition of beautiful basket making. There are many sites around their traditional homelands in Yavapai County which are associated with the Yavapai and Apache people and the entire county is rich with archaeological resources from these and other native cultures.

The town of Mayer, Arizona, lies in the northeastern foothills of the Bradshaw Mountains in Yavapai County, the home of the Yavapai and Apache people. It is within the Big Bug Mining District in central Arizona in a region where deposits of gold, silver, copper, lead and zinc abound. The geology of the district has determined the modern development of the area and its history. However, long before minerals

A Yavapai Indian woman is seated in front of her dwelling, next to a creek, circa 1900. She is wearing the traditional post–Anglo contact clothing of the Yavapai, which some of the women still wear today. Hand Colored Postcard (Sharlot Hall Museum).

An elderly Apache Indian woman is seated next to a burden basket she is weaving, circa 1900 (Sharlot Hall Museum).

1. The Black Canyon Road, or, How Do You Get There from Here?   11

An overview of Mayer circa 1904, looking southeast. The Treadwell Smelter is on the hill to the left and the back and side of the White House Hotel are on the far right. The Mayer Business Block and the Mayer Hotel are in the center. Mayer & Mayer Postcard (Sharlot Hall Museum).

brought modern Anglo settlement to the Big Bug Mining District, there was a rich representation of human activity beyond the habitation by the native peoples in the area from the earliest Euro-American contact forward.

As a result of the Spanish *entrada* in 1582-1583, native peoples, likely the Yavapai, encountered non-native peoples for the first time. Don Antonio de Espejo of Spain financed and led an expedition into what was to later become the Arizona Territory on behalf of the Spanish government. The expedition traveled in the northeastern part of Arizona and along the Palatkwapi Trail to the Verde Valley. In 1598, Captain Marcos Farfan de los Godos also visited the Verde Valley. Farfan was followed by Juan de Onate in 1604. In all of these expeditions, they were seeking the gold that they had heard about, but, not finding any in profitable quantities, they lost interest and moved on. It is unlikely that any of these expeditions actually visited the Bradshaw Mountains and their vast mineral resources.

In *Mining the Big Bug: Archaeological Investigations at Twelve Historic Sites between Mayer and Dewey, Yavapai County*, Pat H. Stein and Elizabeth J. Skinner provide an excellent synopsis of the early non-native settlement of the area:

> When Mexico won its independence from Spain in 1821, the area passed from Spanish to Mexican rule. During the Mexican Period (1821–1848), the Santa Fe Trail opened a lively commerce between Mexican Santa Fe and American St. Louis. The trail brought a tide of American merchants who freighted goods between the two cities. With the traders came trappers—"mountain men"—who scoured the rivers of central Arizona in search of beaver. Mountain men known to have ranged to and through the Yavapai

County area include Pauline Weaver, William Wolfskill, and Ewing Young. Their most intense period of activity was in the late 1820s and the early 1830s. As the market for beaver pelts crashed in the mid 1830s, most trapping came to an abrupt end. However, when Mexico ceded much of Arizona to the United States in 1848, it was from this body of knowledgeable mountain men that the U. S. Army recruited guides for the southwestern expeditions.

Most of present Arizona was acquired by the conquest of Mexico by the United States in the Mexican War of 1848 and the subsequent Treaty of Guadalupe Hidalgo. That land stretched from the Rio Grande to the Pacific Ocean and included more than 500,000 square miles. Additional land was acquired later by purchase. The present states of Arizona, California, Nevada, New Mexico and Utah, plus portions of Colorado and Wyoming, were carved from this territory. But before that could occur, the new boundary between the United States and Mexico had to be surveyed and defined. Representatives of the United States and Mexico met in San Diego in 1849 with the goal of starting the boundary line from west to east. Just as this mapping and surveying project was to commence came the California gold discovery, and many of the soldier escorts necessary for the exploration and mapping project deserted for the gold fields. However, by February 1850 the line had been established from the west coast to the confluence of the Gila and Colorado rivers. At the same time, the Territory of New Mexico, which included most of what would become Arizona, was organized. Then bitter boundary disputes between the United States and Mexico commenced, which were eventually resolved partially by the Gadsden Purchase. This purchase of land by the United States government concluded in June of 1854 and added an area about the size of South Carolina to the southern boundary of New Mexico Territory. This important acquisition was the last step in the western expansion of the United States (except for the later purchase of Alaska Territory).

Meanwhile, in an area known as "Tierra Incognita" that would include the part of Arizona Territory which eventually became Yavapai County, a very important event took place. In May 1863, a prospecting party organized by Joseph Reddeford Walker with John W. "Jack" Swilling as their guide, discovered gold along Lynx Creek south of present-day Prescott. Soon a mining camp called "Walker" sprang up. This was the first gold strike in this part of the territory and was substantial enough to provide the promise of economic boon to central Arizona. Up until the discovery of gold in central Arizona, no one really wanted what would become Arizona Territory. Numerous attempts to obtain separate territorial status for Arizona had failed. Most proposals suggested dividing New Mexico Territory approximately in half from east to west, with the northern half becoming New Mexico Territory and the southern half becoming Arizona Territory. Some in Congress suggested that we had one war with Mexico to win the country, and we should have another to make Mexico take it back. Suddenly, however, the unwanted territory was very desirable: there was gold!

Finally, the Thirty-Seventh Congress of the United States separated Arizona and New Mexico Territories from north to south. On February 24, 1863, President Abraham Lincoln signed the Organic Act, making Arizona a separate territory of the United States. And what did those in the New Mexico Territory think of losing approximately half of their land area? According to an article entitled "Arizona" published in the *Rio Abajo Weekly Press* in Albuquerque on March 31, 1863:

> With indignation by some, with indifference by others while others consider that it will be more beneficial than prejudicial.... It is said to be rich in minerals; but those minerals are of no use to us or anybody else as long as they remain in the earth. We know that it has extensive ranges of rich grazing.... Our stock has never ranged in these regions because of the Indians. It has been said that we have been deprived of that rich mineral country; we reply that New Mexico has not, and never will be, one cent poorer on that account. For never having been in enjoyment of its mineral or grazing ranges, our loss is imaginary and not real.

On October 23, 1863, Executive Order No. 27 was issued by Brigadier General James H. Carleton, Commanding Officer of the United States Army located in Santa Fe, New Mexico. The order states, in part:

> The recent discovery of gold near the San Francisco Mountains within the District of Northern Arizona and the flocking thither of many citizens of the United States ... renders it necessary that a small military force should be sent to these new gold fields to preserve order and to give security to life and property in that region until the civil officers of Arizona now en route from the east shall arrive within that territory and shall establish and set in motion the machinery of civil government.

The 1870 population of the entire Territory was less than 10,000, not counting Indians. However, as the miners flocked to Arizona, the United States government established a series of military forts at strategic locations. Among them was Fort Whipple,

Fort Whipple was established in Prescott in 1864 to provide military assistance to the area and to the miners. This overview is from circa 1870 (Sharlot Hall Museum).

which became an important factor in the settlement of Prescott, the Territorial Capital from 1864 to 1867 and again from 1877 to 1889. The establishment of the forts led to the need for transportation corridors for communication, to supply the forts and to transfer troops and livestock from fort to fort. As the Salt River Valley and Phoenix began to come to the forefront as the Territory's principal agricultural region, it became imperative that a better route than the one from Phoenix to Prescott through Wickenburg and Skull Valley be constructed. In 1870, the Army had begun to develop a very primitive wagon road called the Black Canyon Route which in places followed a foot trail and horse track. On October 3, 1870, the editor of the Prescott *Miner* newspaper, John Marion, who was traveling with Colonel Stoneman, Commander of the newly created Military District of Arizona, wrote about the condition of the road through Black Canyon:

> Why, a California packer would not have attempted to drive his pack train over such a mountain. But, it was the best we could do, and on we went, "slow like a snail," over great, rough trap boulders, some of which were as large as an ambulance. Now and then, the animals had to be unhitched and the ambulances pulled up by means of ropes. Oh! It was trying on the nerves. Our poor nerves gave out early in the day, and leaving men and officers to "do their duty, nobly," we crawled to camp.

By September 27, 1873, a meeting had been called in Phoenix by William Helling & Company, a flour mill operator who had a contract to supply the reservation at Camp Verde. The purpose of the meeting was to raise funds to construct a new road through Black Canyon to Phoenix and the Verde Valley. This route would shorten the trip from Phoenix through Prescott and on to Camp Verde by half. Within days, the company set about the project and sometimes had as many as 50 men working on the road.

At that time, road building consisted mainly of moving the biggest boulders off to the side and grubbing out the brush, following the least resistant natural course. Cutting back the banks and grading out the steeper approaches pretty much completed the job. The constant grinding of the wagon wheels accomplished much of the after-the-fact grading. Washouts were numerous, and mud was a never-ending complication in the rainy season. When it wasn't muddy, the wheels of the vehicles and the mules and horses stirred up choking dust. However, on November 1, 1873, the *Weekly Arizona Miner* enthusiastically proclaimed: "the new road from Phoenix in Salt River Valley, to Camp Verde in the Verde Valley is completed. The first wagons have made the trip with great ease.... Cause for congratulations on Salt River and elsewhere." The *Miner* further declared that the opening of this road was second in importance only to the arrival of the telegraph in Phoenix.

In May of 1877 a contract was let to construct a road from Black Canyon on to Prescott. Just two months later, travelers reported it to be in good condition. The Black Canyon Road eventually became one of the most important public transportation routes from Prescott to Phoenix. Today, it has been replaced by Interstate 17 and State Highway 69, which follow much of the same route but have, at the same time, obliterated much of the original road system.

The Black Canyon Road would be a great shock to anyone traveling in the latter decades of the twentieth century. The mostly one-lane road followed the natural

This "mud wagon" is at Val Verde (now Humboldt). The caption of this real photograph postcard reads, "This will give you an idea of the mud in Val Verde." It was mailed from Val Verde in February, 1906.

This section of the Black Canyon Road near Prescott, circa 1910, shows the cut for the roadbed and the fairly well-graded surface (Sharlot Hall Museum).

contour of the land, up and down, winding here and there, with switchbacks, pullouts and "S" curves. There were "ambush" spots where the Indians could easily attack the lone rider or family group. Later, "highwaymen," creatively called "Knights of the Road," plagued travelers along the road, stealing horses, mail, express boxes, money and personal items of value from the stagecoaches and the passengers. The road was probably two to three times as long as it would be if it were a straight line from place to place. The road was rough and grades were steep. Boulders frequently rolled onto the road, blocked travel and occasionally caused injury or death to people or livestock. Ruts were prevalent and often substantial. Washes were un-bridged and the banks of the wash were simply broken down to minimize the grades into and out of the washes. Deer and antelope were common. There was no pavement. But it cut five days off the route from Prescott to Phoenix through Wickenburg, and as time went on, it became more improved and carried more traffic.

In the late 19th century and up until 1911, stagecoaches traversed the Black Canyon Road, which became known as the Black Canyon Stage Line route and later the Black Canyon Highway. The stagecoaches were very important to travel in the Arizona Territory. The route was from Phoenix through what are now New River, Black Canyon City, Mayer and Dewey and on to Prescott. In 1877, the rich Tip Top Mine was discovered. The mine was located in the Bradshaw Mountains about 15

This four-hitch team pulling a stagecoach near Prescott is typical of the times—the 1880s (Sharlot Hall Museum).

The first claim filed at Tip Top was said to be a "tip top prospect," thus the name of the mine and the town. Joe and Sarah Belle Mayer lived there, where Joe had a restaurant, prior to moving "down the mountain" in 1882. Two of their daughters were born at Tip Top (Sharlot Hall Museum).

miles northwest of present-day New River. In order to process the ore, the company built a mill on the Agua Fria River, close to the existing Black Canyon Road, and established the town of Gillette (or Gillett). Gillette was named for D. B. Gillette, Jr., Superintendent of the mine. Suddenly, there was a great demand for transportation from Phoenix and Prescott to Gillette. By March 2, 1878, James D. Monihon of Phoenix advertised in the *Salt River Herald*:

> Buckboard Express. From Phoenix to the Tiptop. Fare $5.00 or round trip $8.00. Express packages taken at liberal rates. Leaves the Post Office at 8 o'clock every other day. Odd days in February. Connecting with saddle animals from mill to mine.

From Prescott, the trip was soon improved with the addition of a stagecoach instead of the extremely uncomfortable buckboards. Newspapers at both ends of the route were pressing for Maricopa and Yavapai Counties to improve the road. Gillette was booming and the community wanted a post office. The time was ripe for a mail contract. The company of Patterson, Caldwell & Company was quickly formed and, in March of 1878, purchased the Prescott-to-Gillette buckboard line and increased the service to three trips a week. One week later, they purchased Monihon's Phoenix to Gillette stage line. These purchases, plus Patterson's ownership of the Plaza Stables in Prescott, gave them the resources they needed for a through route from Prescott to Phoenix.

Stagecoach travel was an improvement over previous public conveyances, but it was no picnic. In *Catch the Stage to Phoenix*, Leland L. Hanchett, Jr., describes the Concord Stage. Its arrival in Prescott was an exciting event which brought out much of the local population to "ooh and aah" over the coach. Hanchett writes, in part:

Gillette (or Gillett) was a stage station on the Black Canyon Road at the junction of the Agua Fria and New rivers. It was named for D. B. Gillette, Jr., the superintendent of the Tip Top Mine, which was about ten miles up the mountain from Gillette. This is the Gillette Mine, circa 1890. Photograph by Mitchell & Baer (Sharlot Hall Museum).

Down to the wheels, the Concords were works of art. With spokes made of seasoned white pine and hubs of elm and black cherry, the wheels could easily withstand the rigors of Southwestern temperature extremes.

The bodies were solidly built and further reinforced with iron straps. Three inch thick leather through-braces cradled the coach as it hung between the front and back ends of the chassis. Their purpose was to protect the horses from sudden stops, not to comfort passengers who deemed the devices part of the cruel and unusual punishment of stagecoach travel.

1. The Black Canyon Road, or, How Do You Get There from Here?

This stagecoach is pulling into Plaza Stables in Prescott in 1878. Prescott was the terminus of many a long, uncomfortable and arduous stagecoach trip up the Black Canyon Road (Sharlot Hall Museum).

The adjustable leather curtains did little to protect the passengers from dust, wind, snow or rain. With interior seats just over four feet wide, and head room less than five feet high, passenger discomfort was a given.

Possibly the other appointments such as padded leather seats, bright paint, exquisite landscape on the doors or interior gold leaf scrollwork, made up in some way for the horribly rough ride ahead.

Stagecoaches require stage stations to change horses and drivers, and, in some cases to rest, feed or lodge the

This newspaper advertisement from the November 16, 1898, issue of the Arizona Weekly Journal-Miner *provides the schedule for the Mayer to Crown King stagecoach trip, which took 5½ to 6½ hours.*

Mayer and Crown King
STAGE LINE

**MAYER FREIGHTING CO.,
Proprietors.**

Leave Mayer Daily, 10:30 a. m.
Arrive Crown King, 5:00 p. m.
Leave Crown King Daily, 6:30 a. m.
Arrive at Mayer, 12:00 Noon.

Livery Teams
AT MAYER.

travelers. Stage stations were typically located no more than thirty miles apart, and, depending on the terrain and the frequency of communities along the route, could be as close together as ten miles. The route from Prescott to Phoenix was about 100 miles. Traveling at four to five miles an hour, the stages frequently traveled around the clock. Starting in the 1870s, there were several stage stations at various times in various locations along the Black Canyon Stage route. Most were named for a nearby geographic features or for the station keeper. Coming from Phoenix, the first station was New River on the New River; next, after passing through Gillette, was Swilling's at Black Canyon; next was Bumble Bee and next came Antelope, after the stream which runs past the station. This station later became Cordes. The next station was Big Bug, on Big Bug Creek, which later became Mayer, followed by Spaulding's Station (Agua Fria Station) and then on to Prescott. In the *Prescott Weekly Miner* of August 15, 1879, an article describes the Agua Fria Station on the Agua Fria River near present-day Dewey as follows:

> Perhaps the finest and most tastefully arranged parlor, of its size, within a thousand miles of Prescott is that one at the Agua Fria Station, arranged, fitted up and decorated by those two beautiful Misses, Ida and Emma Spaulding. The room is detached from the station buildings, is about eight feet square with a portico in front and the interior is a perfect representation of a ladies' parlor richly decorated with pictures hung about the walls, miniature furniture, house plants and everything necessary for convenience and elegance.

Spaulding Station, also known as Agua Fria Station, was located on the Agua Fria River near Calderwood Butte. Captain Calderwood, who ran the station in the 1880s, had a reputation for dispensing hospitality to all, just as Joe Mayer did at Mayer Station (Sharlot Hall Museum).

The station on Big Bug Creek was originally called "Cottonwood" and was located at the intersection of the Black Canyon Road and the Peck Mine Road. In 1879, a new station sprang up 2½ miles south of Cottonwood Station. The keepers of this "Big Bug" station were Snider and Muncey (or Muncy or Munsey), and they had rights to 160 acres and the station. In the 1880 United States Federal Census, they are the only two inhabitants listed for "Big Bug" (not to be confused with the later-established mining town of "Big Bug," located further to the northwest "up" Big Bug Creek). Big Bug Creek was named for the large bugs that were observed there by the early miners in the area. By 1882, William M. Muncey was the owner of the station and the land. On July 13, 1882, Muncey sold out to Joe Mayer for $1,200 in gold. Thus, Big Bug Station was the beginning of the Town of Mayer.

In May of 1895, a freighter named F. R. Stees wrote a lengthy plea to the Yavapai County officials in the *Arizona Weekly Journal-Miner* about the Yavapai County roads:

> It has been a matter of considerable thought on my part, at least when driving over the roads we have, how little interest is taken in this matter.... The conveyances are first tilted on one side and then on the other, and there are some sharp water cuts, pitches or chuck holes, all causing excessive and dangerous wear and tear on conveyances and teams.... It is a wonder that settlers outside the City of Prescott, do not rise in their might and march in a body to the county officers and say: "...it would look more sensible in you to spend some of your surplus cash to make better roads that would give better facilities to increase the wealth that you are all reaping from our freighting business, that we have carried on over the rough roads that you are giving us."

The Mayer Hotel is on the left, with a wagon parked in front and men on the porch; the Mayer Station is on the right in the distance. This is the Black Canyon Road through Mayer about 1897, when the hotel was new but the railroad had not yet arrived (Charlene S. Thorpe-Fry).

Good roads are a good indication of universal enterprise, and in a rough and mountainous section as this, the engineers should use the closest scrutiny to avoid sharp pitches, which are one of the great and worst features in this whole section. The first cost is the heaviest, but in the long run the cheapest.

To all interested I wish to ask can anything be done to give us leveler roads crosswise, and not have our means of conveyances run first on two wheels on one side, then on two on the other side, and often twisting the whole in all possible shapes!

By the teens, the "stage" lines were still operating, but they were now transporting their customers in motorized vehicles rather than by stagecoach. In the July 15, 1916, issue of *Yavapai* magazine, under the headline "Big Road Work Done," it was reported that "a very remarkable labor has been performed by the Supervisors of Yavapai County in getting the roads connecting the principal places of the county into shape." The article further addresses work being done in District No. 2, stating that "a big force of men is now at work rebuilding the old county road. This is known as the Black Canyon Road, and will be of enormous value to the district." Perhaps the County Supervisors were finally paying attention to Mr. Stees' complaints of 1895. It seems that not too much has changed in over 100 years when it comes to the condition of roads.

By the 1920s, the "stagecoach" had turned into a motorized vehicle. Here, the Whipple Stage, with owner Jack Sills, is posed in the middle of Highway 89 in Granite Dells outside Prescott with his "stagecoach" (Sharlot Hall Museum).

By the 1910s, Black Canyon Road had improved quite a bit. This shows the road near Prescott (Sharlot Hall Museum).

"Prescott Junction," the intersection of the Black Canyon Highway (Highway 69) and Highway 89, has been reconfigured several times. This work is being done in the 1950s by the Arizona Highway Department. This intersection has now been completely reconstructed again (Arizona Highway Department).

In 1938, a proposed highway from Phoenix to Prescott began to show up on Arizona maps as State Route 69. In 1954, the Arizona Highway Department improved the Black Canyon Highway significantly. The route was altered, cut, filled and graded; dangerous curves were eliminated; highway signs were added and portions were paved. And the name was officially changed to the Rock Springs-Prescott Highway. Today, the Rock Springs-Prescott Highway has been renamed and replaced by State Highway 69, but many long-time Yavapai County residents still call it "the Black Canyon Road."

2

The Mayer Family at Big Bug Station

Joe Mayer was born Joseph Hoffmire (sometimes spelled Hoffmayer, Hoffmeier or Hoffmeyer), in Olean, New York, the son of Antoine and Marie Therese Hoffmire. Some sources say that he was born in 1846, but his Arizona Death Certificate states that he was born on April 8, 1851. The family spoke only French. At the age of 14, Joseph left his family as the result of a troublesome home life. He changed his name to avoid being found by his domineering and ill-tempered father. The young man worked in a cigar store and then a cracker factory. He made his way west by train, circus wagon and wagon train, wandering through Nebraska, Kansas, Colorado and New Mexico. It took him several years to travel the country. He worked at many jobs, including on the grading line for the construction of the Union Pacific Railroad. He left that job when the line reached Omaha, Nebraska. There he heard of placer mines in the area of Silver City, New Mexico, and headed south.

He soon found that he lacked the skills for mining, so he signed on as a cook's apprentice, assuring him of three square meals a day and a chance to "rub elbows" with the miners and their silver. Joe was very bright and ambitious. He learned English very quickly, learned to cook and bake and was good with figures. He studied the ways of the successful businessmen he became acquainted with, assessed the needs of the "boom towns" and saved his money. He made friends wherever he went because he was honest, reliable and generous.

According to Joe and Sadie's youngest daughter, Winifred (called Winnie) Mayer Thorpe, in her article in the *Journal of Arizona History* entitled "Joe Mayer and His Town," while he was in Silver City Joe caught the eye of Sarah Belle Wilbur. Sarah Belle, known as "Sadie," had red hair and blue eyes and was tiny and curvaceous. She was born in Maine July 25, 1854, and had come west with her parents, Martha Mary Young and Joshua

This cabinet card portrait of Joe Mayer was probably taken around 1900, when he would have been about 50. It has been published many times, but the location of the original is unknown.

The Globe Mine was located in 1873 and "Globe City" (now Globe) was named for the mine. Joe and Sarah Belle Wilbur were married there in 1877, where Joe operated a boarding house (Sharlot Hall Museum).

From Globe, Joe and Sarah (Sadie) moved to Tip Top, which was booming with mining activity. Their first child, Mary Belle (Mamie) was born there in December, 1878. She was reportedly the first baby born in Tip Top (Sharlot Hall Museum).

Wilbur. Soon after their arrival in Denver, Colorado, Joshua Wilbur died of pneumonia. Sarah and her mother then moved to Silver City, New Mexico, to live with Martha's bachelor brother, Hiram Young. Sarah became a teacher and was a director of the privately-owned Silver City School.

Soon, Joe and Sadie were engaged and were planning to marry in Globe, Arizona. However, according to Winnie Mayer Thorpe, upon their arrival they encountered a smallpox epidemic in the town. Sadie had been vaccinated against the disease, but Joe had not. In her typical fashion, Sadie set to work nursing the sick, including her soon-to-be husband, Joe. Her medicines were the juice from canned tomatoes and castor oil. Soon, Joe, who had lost all of his hair while he was sick with smallpox (his face was pockmarked the rest of his life), was well enough to travel. Joe moved to a mining claim at McMillan Camp outside of Globe where he bought a one-half interest in a boarding house. He rented a house and sent for Sadie. They were finally married in Globe on December 11, 1877.

By 1878, Joe and Sadie had moved on to the booming mining camp of Tip Top in the Bradshaw Mountains of central Arizona Territory. Joe bought a restaurant and started a store. Both businesses harkened back to Joe's earlier experience as a cook's apprentice and the days when he was studying the wants and needs of the residents of the mining towns in which he lived and worked. Two of their daughters were born in Tip Top: Mary Belle (Mamie) in 1878, the first baby born at Tip Top, and Martha Gertrude (Martie) in 1881.

This snapshot of a house in Mayer is similar to a sketch of the "first house in Mayer," drawn by Jack Coker, a friend of the Mayer family. The Big Bug Station probably looked pretty similar to this when Joe and Sadie bought it in 1882.

An article in the *Weekly Arizona Miner* entitled "Letter from Tiptop," dated January 3, 1879, mentions Mamie's birth, stating: "Mr. Joe Mayer, the popular *restaurateur*, was presented early Christmas morning, with a beautiful little girl baby, the first born at Tiptop camp. Joe was, in consequence, much elated over his Christmas and treated the unfortunate bachelors with such an air of superiority that all resolved ere Santa Claus made his annual trip again, they should also be tied in Hyman's rosy chains."

But by 1881, the silver of the Tip Top Mine was beginning to play out, and Joe wanted to find a more promising place to raise a family. He decided to start with Prescott, the home of Fort Whipple and the county seat of the vast Yavapai County. Prescott was growing and was turning into a real city with a spur line, the Santa Fe, Prescott & Phoenix Railway, from the Atlantic & Pacific to the north, with mercantile stores, restaurants, schools and churches. So, he headed for Prescott to investigate. The rude Black Canyon Road took him down the mountain from Tip Top through "exceedingly wild and rough country that even then was recognized as fraught with danger from possible raiding Apaches" to Big Bug Station.

Joe Mayer knew the proprietor of the Big Bug Station, William M. Muncey.

Big Bug Creek was often prone to flooding. This shows the creek in flood in 1898, seven years after the Mayer Station was destroyed in a flood in March 1891 (Sharlot Hall Museum).

Muncey and his partner, Snider, had established the Big Bug Stage Stop in 1879. Muncey and Snider are the only two people listed at Big Bug in the 1880 United States Federal Census. By 1882, Muncey was the sole proprietor of the station and the 160 acres along Big Bug Creek. When Joe Mayer arrived there in the spring of 1882* on his way to Prescott, he was very impressed with the location and the possibilities of establishing a home there. There wasn't much to the Big Bug Station—a "rude little stage station" and a "rough little hotel and store," a corral, a grove of tall cottonwood trees and a clear-flowing stream. Joe Mayer and Muncey, who wanted to return to the California gold fields, discussed the situation. Joe made Muncey an offer, which was accepted, and Joe Mayer left immediately for Tip Top to collect his family and household goods. He paid for the 160-acre property and the stage station on his return trip with $1,200 in gold, which was concealed in his wife, Sadie's sewing bag which also contained thread, needles, scissors and socks to be darned.

As soon as Joe and his family arrived at the stage stop, Joe got busy building a new home and stage station. His three buildings included the family home and guest rooms, which were divided by a large hallway; a mercantile store with an adjacent large dining room, a small bar, and a kitchen; and a barn with a corral.

Joe bought horses and cattle, and soon had a nice herd. He branded the horses with a "BM" brand and the cattle bore a pine tree brand. He planted fruit trees and a large garden, which were irrigated with water from the creek. He dug a well for domestic water.

Although several sources state that Big Bug Station was destroyed in a flood on February 21, 1890, when the Walnut Grove Dam broke during a heavy rainstorm, this

---

*Note: Many sources state that this occurred in the spring of 1881, however, the Mayer's second daughter, Martha Gertrude (Martie) was born in Tip Top in November of 1881, making this earlier date unlikely. Further, the deed for this transaction confirms the date of sale as July 13, 1882.

is impossible. The Walnut Grove Dam was on Hassayampa Creek and would not have flooded Big Bug Creek. Instead, a very brief news report in the *Arizona Republican* of March 1, 1891, reports that "Joe Mayer, who keeps the Big Bug Station and has quite a large stock of goods, had his house, corrals and stable washed away, during the recent storm. He saved his goods, however, and has a force of fifteen men rebuilding." According to Joe and Sadie's daughter Winnie, with the help of neighbors and his employees Joe had moved his family and most of his possessions, including a mother cat and her three kittens, plus some of his store merchandise, to higher ground when the rain started. When the water came roaring down Big Bug Creek, the Mayer family was "high and dry." On March 11, 1891, the *Arizona Weekly Journal-Miner* reported:

> Joseph Mayer, with his characteristic energy and enterprise, has rebuilt a temporary station from the wreck of his former place, and has ordered material for permanent buildings. He saved all of his stock of merchandise and household furniture. The greatest loss sustained by him was the washing away of his fine orchard, grape vines, and beautiful shade trees which represented years of culture and growth, and which cannot be replaced in a few weeks or months as the buildings that were washed away can. Travelers on the Black Canyon Road, however, will be provided, as heretofore, with the best accommodations furnished by any stage station in Arizona.

As soon as the flood was over, Joe started rebuilding his home and the station close to Big Bug Creek, but at a different location. He recycled some lumber from the buildings which were washed down the creek by the flood, and the rest of the building materials arrived by mule-drawn freight wagons. Joe hired carpenters Fred Stees (or Steece, a minister in Mayer) and George Rich to build the new stage station and mercantile store. An article in the *Arizona Weekly Journal-Miner* on March 25, 1891, states that "Joe Mayer has his new hotel about completed on Big Bug, and will give a 'house warming' and dance soon, to which he extends a general invitation to all." The new building was large and spacious and included the mercantile store and a dining room plus living quarters for the Mayer family. Joe hired three or four Chinese boys to run the restaurant. Joe added guest cottages across the street, a barn and a livery operation, and later at Sadie's insistence, he built a powder house in which to store the dynamite about a mile out of town.

The Mayer Station was rebuilt after the flood of 1891. This early snapshot of the Mayer Station shows the north side of the building and one of the bay windows. The woman in the photograph may be Sadie Mayer. There is one dog on the bench and the other dog, which is probably barking at the photographer, is tied up.

Joe Mayer also funded the building of the first school so the local children could walk there. He had one of his livery employees drive some of the local children who lived further away to and from school, picking them up in a light wagon. Some children who lived in more remote areas around Mayer stayed with the Mayer family during school so there would be enough children to "make up a school." The first teacher in the Mayer School was Miss Frances (Fanny) Willard, who later married rancher and Yavapai County Sheriff John Munds. She was the first woman elected to the Arizona Legislature just two years after statehood, in 1914.

The population of Yavapai County in 1870 was 2,142. By 1890, it had grown to 8,685 people spread over a land area of over five million acres. Many of these people came to Arizona and to Yavapai County to "strike it rich" with a mining claim or two. By 1893, more than 24,000 mining claims had been filed in Yavapai County, which was rich in gold, silver, copper, lead and zinc. That equates to approximately three mining claims per man, woman and child. Joe Mayer was no stranger to the mining business, and although he felt he was not "cut out" for the hard work of mining, he was certainly not adverse to becoming financially involved in mining enterprises and grubstaked many a "sourdough." Some struck it rich and paid him back. Some were never heard from again. But Joe never stopped listening to the miner's hard luck stories and provided a stake for almost every miner who came along. In the meantime, Joe was acquiring and purchasing mines and shares in mines in the area and established mining claims all over his property. He became partners in a number of mining operations, including an onyx mine, which he owned for a time in partnership with William Owen "Buckey" O'Neill of Prescott and others.

A sweet portrait of Joe and Sadie Mayer's three children with Mary Belle (Mamie) on the right, Wilbur Joseph (Burr) in the center and Martha Gertrude (Martie) on the left. Although this cabinet card, which was taken in Phoenix, is dated 1888, it was probably taken about 1885 as the children are too young for the photograph to have been taken in 1888 (Sharlot Hall Museum).

After their move to Mayer, Joe and Sadie had two more children: son Wilbur Joseph (Burr) in 1882 and a third daughter, Winifred (often spelled Winnifred) Lucille (Winnie) in 1892. Sadie's mother, Mary (Martha) Wilbur, taught the children to read and write and "do their numbers." In an interview published in the *Prescott Courier* on June 4, 1981, Winifred Mayer Thorpe talked about growing up in Mayer, saying that "life in Mayer was mostly sleepy" and that "excitement was hard to come by." She further said that the "busy throb of progress had little to do with the games,

2. *The Mayer Family at Big Bug Station* 31

Frances "Fanny" Willard on her horse, the first teacher in the Mayer School. She later married Sheriff Johnny Munds of Flagstaff.

This hand-colored postcard is entitled "A Prospector's Outfit" and shows the typical 19th century "sourdough" with his string of burros, sluice box and supplies all neatly packed and ready to go to the next bonanza. Brisley Drug Company, Prescott, Arizona postcard, 1898.

This view of Joe Mayer's Onyx Mine shows a group of men with horses and a wagon, apparently getting ready to load pieces of onyx on the wagon, perhaps around 1900.

This cabinet card portrait of Joe and Sadie Mayer and their son Wilbur was probably taken about 1886. Joe would have been about 35, Sadie 32 and Burr about four. At this time, they would have been living in Mayer at the first Mayer Station. Their last child, Winifred, was not born until 1892 (courtesy Arizona Historical Society/Tucson, no. 61051).

diversions and chores of its children." Winnie's sanctuary was the kitchen, as it was often a matter of the children staying out of the way of the hubbub of the Mayer household. Her chores included locking up the chickens as the "skunks were bad." Winnie said that "sometimes there was action, sometimes it would be pretty dull, then they'd have a dance to wake them up."

Martha Wilbur lived with her brother, Hiram Young, in Mayer, so the extended Mayer family helped out in many of Joe's businesses. Hiram, who was from Maine, grew wonderful potatoes in his extensive garden, which Joe sold in his store, had his own orchard and also had a herd of cattle, which he branded HY. Martha died in 1914. Hiram passed away in the Mayer's home a few years later.

Sadie, too, had her enterprises. She was appointed Postmaster of Mayer by Postmaster General John Wanamaker in 1884. She ran the post office from a small room in the Mayer's home and was frequently awakened in the middle

This document, dated January 11, 1887, is a receipt for a deposit of $25.50 for S. B. Mayer's Post Office license for the first quarter of 1887. Sarah became the postmistress of Mayer in 1884.

of the night by someone who had come into town and wanted their mail. Sundays were no exception for the delivery of the mail. She was also known and respected in the area for her ability to "doctor" those who needed her assistance. One time, she removed a sliver of metal from the eye of one of the mine employees with a magnet. She later became a notary public.

For some reason, in July of 1893 Joe Mayer advertised his property in Mayer for sale in the *Arizona Republican*. As always, he promoted the attributes of Mayer well, touting the business opportunities and stating "one or two live men can make a fortune

The Chance of a Lifetime.

An Opportunity to Step Into A Reliable and Paying Cash Business In 10 Days.

For ten days only the undersigned offers for sale the Mayer or BIG BUG STATION, situated on the Black Canyon road 27 miles from Prescott and 78 miles from Phoenix. For over eleven years the average annual business has been $30,000, and the net yearly profit has been more than $3,000. Desiring, however, to make a change, I will sell

The Entire Business for Eight Thousand, Five Hundred Dollars, Cash,

Or I will trade for Phoenix improved property, or for a ranch near Phoenix. The property which I offer consists of a store, restaurant Postoffice, lodging house stage barns, corral and outbuildings, the actual cost of the buildings being $6,000. Furniture and goods will be sold separate. The stock of goods will invoice from $3,000 to $4,000 and the furniture of the lodging house, kitchen and dining room will invoice $1,000. **All goes for $8,500.** The station is situated at the foot of the Big Bug mining district, and adjoining the Big Bug Onyx quarries. Several roads lead to the mining camps and to the farming valleys. One or two live men can make a fortune in a short time. Satisfactory reasons for wishing to sell.

ADDRESS **JOE MAYER,** MAYER, ARIZONA.

"Desiring to make a change" in July 1893, Joe Mayer advertised his property in Mayer for sale for "ten days only" for $8,500 or trade for a "Phoenix improved property or for a ranch near Phoenix." The property included a store, restaurant, post office, lodging house, barns, corral and outbuildings, with the furniture and goods to be sold separately. Apparently, there were no takers.

This overview of Mayer was probably taken about 1900. The back of the Mayer Hotel (two story building) is on the left with the Mayer Business Block and the Mayer Owl in the center. The Prescott & Eastern water tank and the depot and yard are to the right. Real photograph postcard.

The Prescott & Eastern (P & E) arrived in Mayer in 1898. This is P & E Junction and the depot about 1898, where the P & E left the Santa Fe, Prescott and Phoenix line at Granite Dells in Prescott. One train is coming in from the P & E heading toward Prescott and the other is heading out on the S F, P & P.

in a short time." He apparently changed his mind or there was no interest, because the Mayer family continued to live there for many years.

One of Joe Mayer's business associates and friends was Frank M. Murphy, a man whose name is synonymous with railroad building in Arizona. Murphy was the "railroad king" of Yavapai County, having brought the Santa Fe, Prescott & Phoenix Railway (SF, P&P) from Ash Fork on the trans-continental Atlantic & Pacific to the north to Prescott in 1893. By March of 1895, the SF, P&P had reached its terminus in Phoenix. Later, it was largely through Joe Mayer's influence that Murphy brought the Prescott & Eastern (P&E) railroad into the Bradshaw Mountains from Prescott to Mayer in 1898. The P&E passed through a diversity of countryside on its way from Prescott to Mayer. It left Prescott at the P&E Junction with the SF, P&P in Granite Dells north of Prescott. From there, it traversed Lonesome Valley (now Prescott Valley) to a siding near the road to Jerome. The P&E then crossed Lynx Creek and headed into the lush valley of the Agua Fria River and the depot at Cherry Creek. Leaving the Agua Fria Valley, the P&E headed into the Big Bug Mining District and followed Big Bug Creek to Huron and on to its terminus at Mayer. In *Ghost Railroads of Central Arizona*, author John Sayre states: "the Prescott and Eastern's contribution to the development of Yavapai County cannot be overstated. It gave direction and impetus to the development of the mining industry in the area south of Prescott. The rail brought in almost as much eastern capital as it shipped out ore."

Other business associates and friends of Joe Mayer's were Professor James Douglas and his son, James S. Douglas, of the United Verde Extension Mine in Jerome. It was primarily through Joe Mayer's friendship with James S. Douglas, who lived in Mayer, that the Phelps-Dodge Company brought water to Mayer. The company built a pipeline water system from Grapevine Springs, about eight miles northwest of Mayer.

By about 1897, Joe Mayer had established a brickyard and began building other buildings to house his various enterprises: the Mayer Hotel (1897) and the Mayer Business Block (1902–1904) and the Mayer Apartments (commonly known as Mayer's "red light" district) in 1907 or 1908. The business block included the Mayer Saloon, the Mayer Mercantile and the Mayer Barber Shop, all owned and managed

This early photograph of the Mayer Hotel, built in 1897 by Joe Mayer, shows the P & E water tank and depot in the left background. The photograph is taken facing southeast along the broad, dirt roadway of the Black Canyon Road.

The Mayer Business Block was built in four phases and completed in 1904. The boxes around the trees are to protect them from being chewed on by the horses tethered out front. Tents set up to the south of the Business Block may mean that there is a circus in town. The Mayer Owl and Lev Nellis' meat market are to the far right. This photograph was taken about 1905.

Although this photograph is not labeled, it is most certainly of Martie (left) and Mamie Mayer. It was taken in Long Beach, California, where Joe sent his family for a year in 1893.

Marjorie Belle Looney, daughter of Martie Mayer Looney and Dr. Robert N. Looney, who were married in 1900, in her goat cart. Marjorie was voted the most beautiful baby in Prescott. Real Photograph Postcard taken about 1920.

by the Mayer family. Joe Mayer had a reputation as a fine host, an excellent chef and a good fellow. His businesses prospered, and gradually the "short" route along the Black Canyon Road from Phoenix to Prescott became more well traveled, as opposed to the "long way" by the Wickenburg Road, partly because of the Mayers' reputation for great hospitality to one and all. In an article in *Yavapai* magazine of March, 1918,

Left: *Although this photograph is not labeled either (apparently, the Mayer family wasn't too concerned with labeling the family photographs), it is most certainly of Winifred Mayer and Tom Thorpe. Thomas E. Thorpe was a World War I veteran. Winnie met him at Brisley's Drugstore in Prescott. They were married in 1919.* **Right:** *Winifred (Winnie) Mayer Thorpe on the front porch of the Mayer Station, also known as the "stage stop" in 1978. (Alvin S. Abrams).*

This photograph of the Mayer Station was taken about 1904. Facing southeast, the Mayer Business Block would be on the same side of the street a block or so further away. This shows the bay window on the north side of the building. The barns were in the rear of the property and the corrals were across the street.

it was stated of Joe Mayer that "no hard luck story was ever told him in vain" and "no hungry man was ever turned away from his door unsatisfied."

In 1893, Joe sent his family to Long Beach, California, for a year as he felt a change of climate would be particularly good for Martie, who was slowly recovering from Typhoid Fever. In her article "Joe Mayer and His Town," Winifred Mayer Thorpe wrote about her sisters, stating that "Mamie and Martie were called 'the two beautiful Mayer girls.' Mamie was small and dainty with large pansy-blue eyes like Mother's. She had thick, wavy, long brown hair. Martie had lovely blue eyes, dark hair and a rose-petal complexion.... She favored father in her looks." Martie married Dr. Robert Nelson Looney at the Mayer's home on June 28, 1900. Dr. Looney was a surgeon who had started a hospital in the nearby mining town of McCabe. He and Martie established their home in Prescott. They had one daughter, Marjorie Belle Looney. Dr. Looney died in 1962 in Prescott. Martie died on March 23, 1967 in Mayer. Mamie never married. She attended the University of Arizona and became secretary to Arizona Governor George W. P. Hunt, later returning to Mayer and becoming the Postmistress of Mayer, just as her mother had been. Confined to a wheelchair for many years, she died in Mayer in 1964. Burr attended the University of Arizona in Tucson and business college in Los Angeles, marrying Sarah Annie Skelton in 1903. They had two sons, Joseph H. and Wilbur Nelson (Nelson). Nelson died in 1952 and Joseph in 1967. Joseph and Nelson's father, Burr, died in December 1955 and their mother Annie died in 1965 in California. Winnie married Thomas Edward Thorpe, a World War I veteran, in July 1919, in Mayer. They had two sons, Thomas Edward Thorpe, Jr., and Wilbur Robert Thorpe. Winnie and Tom Thorpe were divorced in 1937. Winnie lived much of her life in Mayer in the old family "stage stop." She died in Prescott on July 19, 1983, at the age of 90.

In the January 5, 1898, issue of the *Arizona Weekly Journal-Miner*, published in Prescott, an article about Joe Mayer states: "Mr. Mayer located at this point 16 years ago, and as he stated in a conversation with your reporter 'If a man settles down he can do well anywhere,' the fact seemed to us proven true in the case of our host, who, by his business sagacity and stick-to-ativeness that characterize men of positive character, has prospered and made a success of life."

On the cold and rainy night of Saturday, November 27, 1909, Joe Mayer heard noises out in the yard. He went outside with his loaded "44 caliber six-shooter," stumbled or slipped and fell, causing a deep gash in his head and face, and accidentally shot himself in the upper chest, inflicting a mortal wound. Before he died, he assured those in attendance, which included his son-in-law Dr. R. N. Looney and Dr. Clarence Yount, both of Prescott, that it was an accident and that he did not want anyone to "entertain any other reason." He died at 5:00 P.M. at his home in Mayer. An inquest was held in Mayer on November 28 by Martin T. Knapp, Justice of the Peace and Coroner Ex-Officio. The findings were that "Joe Mayer accidentally shot himself while falling." In an article reporting Joe Mayer's death in the November 28, 1909, issue of the *Prescott Journal-Miner*, it was said of him "the past life of this prominent Arizonan will prove interesting, carrying as he did from the day he entered the Territory to the close of his life, a fellowship that will live long after he is laid away."

Sarah B. Mayer lived the rest of her life in the "stage stop" in Mayer. She died on November 11, 1934.

3

The Town of Mayer:
How the West Was Really Won

It was not uncommon in the mid-to-late 19th century for enterprising adventurers to travel west to start a new life. Most of these adventuresome folks were single men, and some were men who left their families at home, although some brought their families with them. The lure of the idea of the untamed and lawless wild, wild West brought some. The lure of gold brought some. The lure of supplying all of those who were lured by the gold brought some. The lure of open grasslands as far as the eye could see brought some. And some, like Joe Mayer, could see the potential of the West and envision his role in making not only a living, but a good life. Mayer certainly accomplished that goal.

When Joe and Sadie Mayer moved to the Big Bug Station, Joe immediately set to work to improve it and make it the best stage station on the Black Canyon Road. To be the "best" at whatever enterprise they attempted would be Joe and Sadie's lifelong goal. He realized more than others the richness of the country around Mayer. He had wonderful dreams of the Mayer of the future.

In the 1890s, the rich mining resources of the Bradshaw Mountains were in desperate need of capital. In order to attract investment in the mines, the local mine owners and capitalists in the eastern United States clamored for a railroad into the Bradshaw Mountains. At the time, only the highest grade ore could be shipped to smelters as far away as Denver or

The Callen Party Wagon train at Junction City, Kansas in 1875. Kansas was the major jumping-off point for the Santa Fe Trail and the West. Photograph by A. P. Trott (Sharlot Hall Museum).

A mounted cowboy watches over a cattle herd along a stream in Yavapai County about 1900. This is typical of the open range land around Mayer in the late 19th and early 20th centuries (Sharlot Hall Museum).

A photograph of gold nuggets found at Big Bug, Arizona (Sharlot Hall Museum).

Bags of ore ready to be shipped the "old fashioned" way, by burro, at the Turyriver Mine in Gleeson in the 1880s. This scene is typical of most small ore producers before the arrival of the railroad (Sharlot Hall Museum).

This street view of "downtown" Mayer along the Black Canyon Road was taken about 1905. It shows the completed Mayer Business Block on the right and the Mayer Hotel on the left. The street trees are still boxed to protect them from the hungry livestock (Sharlot Hall Museum).

The Granite Dells north of Prescott provide a scenic backdrop for a Santa Fe, Prescott & Phoenix locomotive about 1907. This is a Brisley Drug Company hand-colored postcard printed in Germany.

El Paso due to the high cost of shipping. Lower grade ores were simply stockpiled at the mines. Naturally, Joe Mayer wanted that railroad line to go through his town, and he strongly lobbied his friend Frank Murphy, president of the Santa Fe, Prescott & Phoenix Railroad, to terminate the new line at Mayer. Eventually, Joe and others convinced Murphy to route the railroad through Mayer to take advantage of Mayer as a shipping point for minerals, sheep and cattle. As the surveys were being done for the railroad, Joe Mayer, who was anticipating increased business from the railroad, hired architect and builder Hill C. Moore to build a two-story store, hotel and bar near the old stage station. The Mayer Hotel was described in the *Arizona Weekly Journal-Miner* on October 1, 1897, as "one of the best appointed and handsomest buildings in Northern Arizona." It was completed that year. In 1898, the Prescott and Eastern Branch (P&E) of the S F, P & P was completed from the Prescott and Eastern Junction northeast of Prescott to Mayer. The first train arrived in Mayer on October 15, 1898. As John W. Sayre states in *Ghost Railroads of Central Arizona*:

> The Prescott and Eastern Railway was a relatively short railroad but crossed a diversity of beautiful countryside. The first two miles of the line from the P&E Junction passed through the scenic granite landscape of Granite Dells. The rail then passed through Lonesome Valley and on to a siding along the road to Jerome.... The line then crossed Lynx Creek and descended into the lush grain fields of the Agua Fria Valley. Near the old Hildebrandt Ranch in the Upper Agua Fria Valley, the first depot on the line, Cherry Creek, was established. At the Bowers' Ranch, near present day Humboldt, the P & E veered southwest toward the Bradshaw foothills and the Big Bug country. As the road bed ascended from the Agua Fria Valley, it followed Big Bug Creek, past the depot at Huron and on to the terminus at Mayer. The peaks of the Bradshaw Mountains peered through wispy clouds and were visible from the depot in Mayer. They beckoned the rail onward.

This is a view similar to the previous photograph of downtown Mayer. The circa 1907 photograph shows the Mayer Hotel on the left, the Mayer Owl on the far right, and the Mayer Business Block with a couple of freight wagons parked out front and the white fence around the Mayer Station in the center. Joe Mayer's street trees had grown quite a bit.

In an article published in the *Arizona Weekly Journal-Miner* on December 22, 1898, Joe Mayer expressed his feelings about Mayer and the surrounding area: "Joe Mayer had great confidence in the future of the Big Bug Section of country and thinks it will not only support a large and thriving mining population, but be a great back country support for Prescott, if Prescott will put on sufficient progression to keep pace with the back country."

The virtues of Mayer were further elaborately praised in an article subtitled "Like in Pioneer Days, it Remains in the Saddle and has a Bright Future Ahead," which was published in the *Arizona Weekly Journal-Miner* in May 1900. The writer went on to say, in part:

> Probably in all Arizona there is no section more familiarly or agreeably known to the masses than can be said of the old station of Mayer. In the earlier days of staging from Prescott to Phoenix, it was regarded as a luxury when the wild mustangs pulled up at the place, and after leaving it the universal verdict was that in hospitality and accommodations it savored of the delightful and refreshing indeed. So famous had this center become in the annals of overland travel, that the monotony of riding on the hurricane deck of a prairie schooner was more or less of a pleasant anticipation after one had partaken there of the rough and ready generosity, to await him at every turn. With the decline of the stagecoach and the building of the railroad, however, one would naturally believe that the old landmark of the "fast and furious" was to be supplanted by a delicacy of regard that the tenderfoot era was sure to unravel, or, in other words, the place was to lose its old time identity and charm. But such is not the case.... In short, Mayer

of old is still the same, but as the boys say, on a bigger scale.... On the contrary, the railroad has brought hundreds where the stagecoach carried one, and with the change the country in and around that place has been more benefited and placed in better light than could have possibly be accredited to any other channel. With the change, Mr. Mayer has kept pace with the times, and we are pleased also to state as the avenues of mining progress widened, so has he, in a great measure, been benefitted. As an industrial center Mayer is handsomely located. It supplies Big Bug, up and down, while for Bumble Bee, Black Canyon, Agua Fria, the Bradshaws and other districts it is an ideal location as a distribution point.

By 1901, Frank Murphy had the backing to build two branch lines of the P&E to serve the mining districts in the rugged Bradshaw Mountains: one to serve the rich Big Bug Mining District (the Poland Branch) and one to serve the Peck, Tiger and

This panorama or "double" postcard of the Bradshaw Mountains and the Crown King Branch of the Bradshaw Mountain Railroad looking west shows the switchbacks necessary for the locomotives to climb the mountain. The Swastika, De Soto and Oro Blanco Mines are in the photograph as is the "Black Warrior Trestle," the largest trestle on the line.

This overview of Mayer would have been taken about 1910. The Treadwell house is in the foreground. The large trees curving off to the left follow Big Bug Creek. The 1902 Schoolhouse built by Joe Mayer is in the center, just northwest of the railroad tracks with the water tank, depot and railroad buildings beyond. The back of the Mayer Hotel can easily be seen in the center of the photograph. To the left and above the town is the Treadwell Smelter with its substantial siding and trestle. The Jagged Tooth Mountains are in the far background.

Pine Grove Mining Districts to the south (the Crown King Branch). The Bradshaw Mountain Railroad was partially constructed from rail which was removed from an abandoned section of the S F, P & P. The materials and manpower were stockpiled at Mayer, another boon for the enterprising Joe Mayer. For convenience, stockyards soon occupied the center of town along the main railroad spur just south of downtown and west of the railroad tracks. Thousands of head of cattle and sheep were shipped, causing the local residents to complain about the odors and the mess.

In 1902, Joe Mayer built the first phase of a new brick business block across the street from the Mayer Hotel which included a saloon, restaurant, large mercantile store and a combination bath house and barber shop. He also built two small apartment buildings in 1907 near the hotel to house the overflow from the hotel. Next, he donated funds to build a new schoolhouse, which opened for 15 students in 1902. That same year, he persuaded the Phelps Dodge Company to bring in a water supply system from Grapevine Creek

The Mayer stockyards were moved into town after the arrival of the railroad in 1898. This photograph of the shipping pens in Mayer may have been taken in the late 1930s.

The Mayer Laundry was located across the street from the White House Hotel. The building went through a number of uses and has now been demolished.

about eight miles away. Joe Mayer was also one of the Territory's first land developers. He, along with others, platted the Town of Mayer in 1904 and promoted the area as a healthy, dry climate and a retreat from the big cities of the East. Lot prices ranged from $100 to $500 each. The area around Prescott was already well known as a haven for those suffering from respiratory diseases and ailments, partially due to the worldwide reputation of Dr. John Flinn, a Prescott doctor from Switzerland who was a pioneering expert in the treatment of tuberculosis. Mayer offered a similar, though somewhat milder, climate than Prescott. It was Joe Mayer's promotion of the area that brought many newcomers to the area in the early 20th century.

Joe Mayer wasn't the only entrepreneur in town. Starting in about 1890, Mayer's town became a center for the outlying ranchers and miners to get supplies, and other adventurous folks took the chance and established businesses in Mayer. Although Prescott was the supply center for the western side of Yavapai County, it was 25 miles from Mayer, and until the railroad arrived, it was a long and arduous trip requiring several days' travel to do business in Prescott.

Joe Mayer and others began to develop all types of businesses. In 1899, Sam Lee, an "Oriental American," established the American Laundry in Mayer. Most had some relationship to mining, which was the major industry of the area. The 1900 United States Federal Census gives the population of Yavapai County as 13,799. The area included in the census for Mayer is not defined, and since Mayer was, and still is, unincorporated, and its boundaries were not specific, the exact area which is included is not clear. The census for Mayer lists a variety of professions, many of which are directly related to mining, such as "prospector" (14), "quartz miner" (15), "mine superintendent" (1), "mine foreman" (2), "machinist" (1), and "assayer" (1). Other occupations

These cowboys are camping out and cooking on an open fire about 1909. The wagon in the right background appears to be the "chuckwagon."

The location of this blacksmith's shop in Mayer is unknown. The man standing on the left is probably Joe Mayer. It is difficult to date this photograph. It could be anytime between the late 1880s to the early 1900s (Garry Cooper).

Mayer Blacksmith Shop
MAYER, ARIZONA.
All Kinds of Iron and Wood Work, Mine Windlasses. Buckets, Drills, Picks, Etc. . . . . .

HORSE SHOEING
New Wagons Built, Second Hand Wagons For Sale. Also a Few No. 1 Mountain Spring Wagons and Buckboards of our own make which we guarantee.

WORK GUARANTEED and PATRONAGE SOLICITED

This advertisement for the Mayer Blacksmith Shop ran in a 1904 issue of the Arizona Weekly Journal-Miner. *The proprietor advertised "all kinds of iron and woodwork, mine windlasses, buckets, drill, picks, etc."*

represented in Mayer at the time included five cowboys; seven "stock raisers," two of whom were raising goats; three blacksmiths; six teamsters; one railroad station agent; and one railroad engineer. There were also one each day laborer, teacher, barber, stenographer, store salesman and carpenter. There was one hotel, the Mayer Hotel owned by Joe Mayer, and there were five cooks. The Mayer family had four children in school and four Chinese servants. There were nine boarders in their hotel at the time of the census. However, much of the area population was located at the surrounding mines and ranches. Even though it was much easier to get to Prescott in 1900 than it was in

A photograph of the Mayer Hotel taken in the 1940s or early 1950s shows the deterioration of the brickwork under the porch. It's no wonder a new set of stairs were built on the side of the porch as the front stairs appear to be a disaster waiting to happen. There is a faded sign painted on the brick which says, in part, "sodas." By this time the Mayer Hotel had been converted to apartments for some time.

pre-railroad days, these nearby mines and ranches were primarily supplied at this time by the merchants and residents of Mayer. Mayer was definitely growing up. Mayer was small in population but, as a shipping point for ore, cattle and sheep and general merchandise, the community was very important to the area as a mercantile and industrial center. Telephones arrived in the area in 1902, and the Prescott telephone directory of that year listed two businesses in Mayer—Grant Bros. and Joseph Mayer's Store, both of which presumably had a telephone with a requirement that the caller "call for these stations by name." The 1907 Prescott area telephone directory listed six residences and about 12 businesses, including the P&E Depot, the Wagner and the Mayer hotels, the Mayer Hospital and several mining companies, all "long distance stations."

By the 1910 United States Federal Census, Yavapai County's population had grown to 15,996. Mayer was growing, too. The 1910 United States Federal Census' list of occupations shows how much more "civilized" Mayer had become since the turn of the 20th century, including house carpenters (7), music teachers (2), several butchers, a seamstress, a stenographer (Mary B. "Mamie" Mayer), a paper hanger, a bookkeeper, a barber, a laundryman and a cobbler. Several restaurants helped feed the citizens and provided jobs, including one owned by a Chinese man, Charley Kung (or Fong). Mining occupations still were a major element of Mayer's economy, with 29 miners in town, along with the president of a gold mine, two superintendents, two foremen, one manager, three mining engineers, a "furnace man" and five machinists. Others worked as laborers (13), woodpackers (1), stockmen (2) and shepherds (5), farmers (1) and "truck farmers" (3). Three mercantile stores, including the Mayer & Mayer mercantile, which was being run by Joe and Sadie Mayer's only son Wilbur "Burr" Mayer and his wife Annie, a lumber store and a hardware store helped supply the community and the miners. The P&E provided jobs for a contractor, section foreman, station agent, two brakemen and six laborers. There were still three blacksmiths, one livery stable owner, one liveryman and three teamsters in Mayer. One man was listed as a "driver," presumably of a motorized vehicle. Dr. Riley Shrum worked at the Mayer Hospital.

This undated portrait of Mamie Mayer appears to show her in her thirties, which would be about 1910–1915. Mamie had been a secretary for Arizona Governor George W. P. Hunt. In 1910 she was back in Mayer and was listed in the census as a "stenographer." Mamie took over the postmistress job from her mother about this time, while her mother continued to help out in the post office.

Another unlabeled and undated photograph, this is certainly Wilbur "Burr" Mayer and his wife, Annie Skelton Mayer. From the clothing and the car, it appears to have been taken in the late 1910s or the early 1920s.

The mines consumed an enormous amount of wood and most households and businesses were heated with wood or coal. The woodcutters and woodpackers were kept very busy cutting, stacking and delivering wood. This is Mr. Yoeman's wood yard at Lynx Creek about 1890 (Sharlot Hall Museum).

The 1910 census also included a special Indian Census. Indian families had long lived in the Mayer area, and Joe Mayer admired and befriended many of them. One of their encampments was among the trees on the opposite side of Big Bug Creek not far from the Mayer's home and stage station. These members of the Yavapai Indian Tribe were moved to the reservation at San Carlos in 1893. However, before that time, Joe Mayer had traded with them and acquired quite a number of the beautiful baskets which were made by the women. These baskets were still with the Mayer family when Winifred Mayer Thorpe died in 1983. In her

Although the Yavapai Indians had been moved to the reservation in 1893, by the time this photograph was taken, perhaps around 1910, they had returned to their homelands in the Mayer area. Here, they are waiting on the platform of the Mayer Depot (courtesy Arizona Historical Society/Tucson no. 49382).

1978 article for the *Arizona Journal of History* entitled "Joe Mayer and His Town," Winifred wrote about the Yavapai who lived nearby (she referred to the Indians as "Apache"): " Often at night when the Indians danced, the townspeople would gather to watch them. It was a colorful sight. They were accompanied by the beating of drums and their weird musical chants, with painted bodies moving in slow grace. Seen by the light of the campfires, it was something one did not soon forget." The special Indian Census, although only three pages, is very telling about the situation of the Indians at the time. Of 55 entries, 12 spoke English and six had been to Indian boarding schools and could read and write English. All but three families had Anglo-Saxon names. Three families had Indian names and a few individuals had nicknames, such as "Indian Jack." There were 17 women listed as basket makers. Most of the men worked as laborers on the railroad or doing "odd jobs" or on farms or as woodchoppers. Many were unemployed and had been for a long time. All of the Indians enumerated on this census were born

Joe Mayer admired the Yavapai women for their basketmaking skills and acquired quite a collection of their baskets. This Yavapai woman is standing in front of a traditional brush shelter about 1900. She is carrying a water jar by a trumpline around her forehead, a traditional way of carrying all sorts of things in containers such as jars or baskets, which leaves both hands free. She is also holding a basket and what is apparently the material for making baskets (Sharlot Hall Museum).

in Arizona and all of their parents were born in Arizona. They were all 100 percent Indian, with one exception—a 20 year-old woman who had attended boarding school, read, wrote and spoke English and was one-half Indian and one-half Negro. Their location was listed as "Mayer Town, Prescott Station Forest Reserve."

By 1913, the Yavapai County Directory had 42 listings for Mayer and very conveniently listed each person's business or work affiliation. Neither addresses nor telephone numbers were listed—everyone in town knew where everyone else lived and where the businesses were located, so there wasn't any need for that information. Sadie B. Mayer remained the postmistress. Joe Mayer was still listed, although he had died four years earlier. Businesses included Arctic Ice and Meat Company, the Arizona

An invoice from one of the Mayer lumber stores, W. A. Frazer, to rancher Fred Dugas, dated August 1917. Dugas Ranch is about 20 miles as the crow flies southeast of Mayer, but about 40 miles by road. It was founded by the Dugas family in the 19th century (Larry Howard).

Another invoice for Fred Dugas, this time from W. S. Goldthwaite, which sold lumber, mining timbers, paints, oil and building supplies, dated in October of 1922. A 16 foot 2 × 4 cost about $1.25. Typical of the times, most regular customers had accounts which they paid on when they could. In this case, the interest on an account over 30 days old was ten percent (Larry Howard).

Here, Yavapai women in traditional dress in 1903 are carrying, according to the caption on the photograph, water and wood. The water tank for the P & E is in the background. Downtown Mayer is to the left.

Power Company, the Mayer Hotel, which was then owned by the Justice of the Peace, M. T. Knapp; Mayer & Mayer, the Mayer Water Company, the *Mayer Miner* newspaper, James E. Hill, proprietor, the Rigby Mining and Reduction Works, Mary L. Wells, grocer and Yavapai Supply Company (hardware), plus a barber, a shoemaker and several saloons.

The March 10, 1898 issue of the *Arizona Republican*, published in Phoenix, had the following to say about Yavapai County and its early residents:

> It is a source of no little pride with the residents of Yavapai County that we include more men whose names are linked with Arizona history than any other county in the territory and no names are more familiar in that connection than those of J. L. Marr, Uncle Dick Thomas, J. K. Hall, Joe Mayer and other residents of the Agua Fria Valley and vicinity. It is certainly to be hoped that nothing will drive such men from the territory. They are too good men for any country to lose.... It is a wholesome education for a young man to live among and learn to know some of these old Hassayampers.* What they have done and what they are doing makes a youngster feel that he must keep hustling to be in it at all.

And that is how the West was really won.

---

*A "Hassayamper" can be defined as a pioneer of pioneers, an early settler in Yavapai County.

4

Ranchers and Farmers Follow the Dusty Trail to Mayer

"Life ain't in holding a good hand—but in playing a poor hand well."—(Epitaph quoted by Sharlot Hall)

As Bert Fireman wrote in his article "Gravel in the Barley," which was published in *Echoes of the Past, Tales of Old Yavapai Arizona,* Volume 1, "Nobody knows when agriculture started in Arizona. The oldest discovered evidence of man's residence under our turquoise sky includes relics that show that the men who built these ancient habitations also cultivated the soil. Arizona's earliest known residents were not hunters alone but raised crops along the streams that, even today, thousands of years later, furnish the life-blood of Arizona's farming industry."

Sharlot Mabridth Hall grew up on Orchard Ranch, near Mayer, with her parents and brother Edward. This photograph shows Orchard Ranch about 1890 with the fruit orchard in the foreground (Sharlot Hall Museum).

4. Ranchers and Farmers Follow the Dusty Trail to Mayer

The Great Seal of the state of Arizona incorporates into its design the "Five C's" of Arizona: Copper, Cattle, Cotton, Citrus and Climate. These "Five C's" represent the economic engines of the State of Arizona from the days of statehood in 1912 to the present—mining, ranching, agriculture and tourism. Although many newer industries and businesses have come to Arizona over the last 100 years, the industries of mining, ranching, agriculture and tourism are still mainstays of Arizona's culture and economy. In Yavapai County, mining and ranching are still very viable industries, as is tourism. Farming in Yavapai County has not been sustained to the extent it was in the past, but it is still a factor, with corn, hay and alfalfa being the most common commercial crops. Fruit is still grown in some areas of the county, but large commercial orchards are a thing of the past. However, remnants of the orchards of long ago can still be found in some very remote areas of the county and are always a delightful surprise when discovered. Further,

Sharlot Hall, poetess, writer and the first Arizona Territorial Historian, founded Sharlot Hall Museum in 1927. She is photographed here about 1905 in her office at Orchard Ranch, surrounded by the books and artifacts that she loved so much (Sharlot Hall Museum).

Orchard Ranch was pretty isolated and when Sharlot Hall went to "town" it was usually to Prescott, an all day trip each way. Here, Orchard Ranch was photographed in the winter when the "snows often covered the grassy plain" about 1890 (Nancy Kirkpatrick Wright in **Sharlot Herself***) (Sharlot Hall Museum).*

Sharlot Hall wrote of the loneliness of Orchard Ranch and Lonesome Valley. As Nancy Kirkpatrick Wright wrote of Lonesome Valley: "An immense valley stretches south from Seligman past Prescott and almost to Humboldt, Arizona." Today, Sharlot would not be lonesome as much of Lonesome Valley is filled with the Town of Prescott Valley and its 35,000 plus residents (Sharlot Hall Museum).

In 1941 Rosena Promberger Minucci photographed the ranch house at the Lessard Ranch on the outskirts of Mayer.

ranching is still a major economic engine, but many of the old farms and ranches, especially those close to major highways and cities and towns, have been broken up and sold for development.

It is impossible to write about ranching and farming in Yavapai County without writing about Sharlot Mabridth Hall. Not only was she a rancher and a farmer herself, but she wrote extensively about her experiences through her essays and poetry. In her

4. Ranchers and Farmers Follow the Dusty Trail to Mayer

After Sharlot Hall's death in 1943, Orchard Ranch fell to ruin. Eventually, the Fain family, area ranchers, acquired the property. This photograph, dated in 1969, shows the ranch house shortly before it was demolished. The location is now the site of the Rafters 11 RV Park. There is a very nice historical display at Rafters 11 about the Hall family and Orchard Ranch (Sharlot Hall Museum).

Poems of a Ranch Woman, collected and published after her death, she told the poignant, often tragic and sometimes humorous stories of ranch life in Yavapai County in the late 19th and early 20th centuries. Sharlot Hall came to Arizona in 1881, when she was 11 years old, with her parents, James and Adelaide Hall. After a couple of years of unsuccessful attempts at placer mining, the Halls settled on the 160-acre Orchard Ranch with Mayer as the nearest town (the site of the ranch is now a part of the Town of Dewey-Humboldt). James Hall raised horses and did a "little of this and little of that." In *Sharlot Herself*, author Nancy Kirkpatrick Wright described the setting of Orchard Ranch:

> Orchard Ranch lay on the edge of Lonesome Valley and Orchard Ranch was Sharlot's home for almost forty years. The ranch was about five miles from the ranching/farming community of Dewey and perhaps fifteen miles by winding wagon road east of Prescott. The area was sparsely populated in the 1900s, growing to no more than 50 families in the 1930s. Pictures of Orchard Ranch show a two-story board and batten house surrounded by a spiked fence and then a sprinkling of chaparral vegetation on the surrounding hills. A windmill rises starkly at the rear of the house, with a few outbuildings and trees barely reviving the harshness.

A Yavapai County farmer is shown checking his crop, which is surrounded by fruit trees, in this circa 1900 photograph. Based on the terrain, this farm is probably located in the foothills of the Bradshaw Mountains.

Sharlot described Lonesome Valley as "bare and brown and lonesome as its name implies—and as beautiful as only the southwest plains can be." She mentioned in her diary the isolation she felt there, especially during the winter months. Although her poem "Old Governor's House" is about the Governor's Mansion in Prescott, an excerpt from the poem could just as easily apply to Orchard Ranch, which was torn down after Sharlot's death:

> Life and death and love and fear–
> Each one has its moment here.
> Now those first come men are gone
> And the old house stands alone.
> Filled with whispering memories,
> Haunting, half hearted melodies–
> A shrine beside the busy way
> To hold the Soul of Yesterday.

The ranchers and farmers who settled Lonesome Valley and the areas at the foot of the Bradshaws around Mayer, such as George E. and Angie Brown, who lived on a ranch along the Agua Fria River, and those who lived along Antelope Creek, Big Bug Creek and Turkey Creek, were all seeking the same things: room, independence and a living. Ranchers raised livestock, mostly cattle and sheep, but also horses and Angora goats, to sell on the hoof or to shear for their wool or hair. Farmers of the late 19th and early 20th centuries, whether growing fruit, vegetables, alfalfa or hay, were raising not only crops to sell, but crops to sustain themselves and their families.

A booklet published in 1907 by the Yavapai Commercial Club, a "booster" group in Prescott, entitled "Yavapai County Arizona the Treasure Vault of the Southwest" states that the purpose of the booklet is "to better acquaint the capitalist, the investor, the homeseeker, the artisan, the laborer and the world at large with the conditions as they exist and the great opportunities offered in Arizona, particularly that district tributary to Prescott." In the section "Agriculture and Horticulture," the author promotes the fact that there is ample land and a ready market in the mining towns available for agriculture and states that "we have room for many thousands of substantial citizens, who will engage in ranching and farming in its various phases, to large profit for themselves, and with much satisfaction to the chief industry of mining."

In the late 19th and early 20th centuries, almost every ranch and farm had a truck garden. Everything possible was grown at home and canned or dried for storage. To supplement the homegrown produce, hunting provided meat—fowl, rabbit, deer and antelope. Cattle, hogs, chickens and turkeys were raised for meat for the family. Cows and chickens provided milk and eggs. Meat, eggs and milk could be stored in the summer in a homemade "desert cooler" for a few days. The excess meat was dried into jerky. Extra milk became butter, and eggs were sold or bartered. Many ranchers and farmers strived to be as self-sufficient as possible, going to "town" only when necessary and often bartering their produce or livestock for the goods they could not provide on the ranch. An article in the October 15, 1916, issue of *Yavapai* magazine reported that Mr. D. W. Thomas "has developed a very pretty place on the upper Big Bug, 6 miles from Humboldt. He recently brought in a wagon load of peaches to Prescott. Mr. Thomas has won numerous prizes at the Northern Arizona Fair for his fine apples as well as fine peaches."

Hunting was a popular sport and a way to put additional food on the table. Here, a group of hunters from the Mayer area have had a successful bird hunting trip, which was probably helped by their hunting dog. The photograph was probably taken around 1920.

This panorama of Mayer, circa 1910, shows the Treadwell Smelter on the far right and the town spread out along the creek. The Mayer Hotel, the railroad tracks with the P & E water tank, depot, and warehouses are on the left center. The train yard is full of Pullmans, boxcars, cattle cars and gondolas and an engine is approaching from the south. The stockyards and loading chutes are to the southwest of the depot and warehouses. The Rigby Reduction Works are on the hill above the town to the left. Photograph by Aveldson Studio, Jerome, Arizona (The Coffeegram, Mayer, Arizona).

Mayer was also a great spot for farming. In Mayer, Joe Mayer had an orchard and a small herd of horses with the BM brand and a herd of cattle which bore the Pine Tree brand. Sadie Mayer's mother, Martha Wilbur, and Martha's brother, Hiram Young, had a substantial garden in Mayer. Joe and Sadie's daughter, Winnie, in her article in the *Journal of Arizona History*, writes that "before long Hiram took advantage of the rich soil and started his own garden. He set out an orchard and bought cattle which he branded HY. His garden was very productive, his potatoes being exceptionally good. Joe bought them and sold them in his store. Hiram used to say 'I know how to raise the best potatoes because I come from Maine.'" In the booklet published by the Yavapai Commercial Club in 1907, the author wrote about Mayer, stating that "there are numerous little valleys adjacent to this town, that could be farmed to advantage where farming and the raising of livestock could be followed with profit. All kinds of garden truck, fruit, etc., find ready local sale at high prices."

A 1923 article in the *Arizona Farmer* ran an article headlined "Mayer a Garden Spot Offers Opportunities." The article goes on to state:

> Mayer is a little garden spot on the Black Canyon highway. Not only are there beautiful ranches about the town, but in the little town itself are many miniature fruit ranches. Among these later are two that are worth mentioning—the Heffleman place and the Slak place. Here are many fine trees in full bearing, apple, crab-apple, all kinds of peaches, pears, cherries, plums and quinces. The Slaks raise many grapes and some Himalaya berries. There are few fig trees in the town, the fig requiring a warmer climate, though out at the Jim Cook ranch many fine figs are raised. Out on the Stewart and old Alexander ranches they grow wonderful apples. Melvin Todd also raises fine apples and peaches, but his cantaloupes are the finest in the valley. Lazard's ranch also produces a great amount of fine fruit, one pear tree bringing in more than thirty dollars last year.

A subsequent article in the *Arizona Farmer* two weeks later exclaimed that "imported" strawberries were selling for 30 or 35 cents for a small box, but that in Mayer, there were three little "strawberry patches" which furnished strawberries, and plenty of them, for the entire summer and into the fall. The article suggested that strawberries would be a good commercial crop in spite of the fact that "you can't give a strawberry bed too much water." Today, there is no commercial farming in Mayer, but there is commercial farming in the area and in the nearby Town of Dewey-Humboldt. Lonesome Valley and the foothills of the Bradshaws are home to many beautiful and

4. Ranchers and Farmers Follow the Dusty Trail to Mayer

A group of fourteen mounted cowboys are "ready to start" in this hand-colored postcard from the Mayer area. The caption on the reverse of the postcard reads: "Singly or in two or threes they gallop down the road on their wiry little horses, their light, supple figures erect or swaying slightly as they sit loosely in the saddle," a very romantic and not very realistic description of the hard-working cowboy. The postcard is dated November 12, 1907, and was mailed from Mayer to Tweed, Ontario, Canada.

productive home gardens along with a number of small, successful, family-run commercial farms which provide some of their wonderful produce and flowers at the local farmers' markets.

Cattle raising in the Arizona Territory really began to become a business with business-like plans about 1880. As thousands of longhorn cattle were driven west from Texas, either by trail or by rail, Arizona and New Mexico Territories, with their luxurious grasslands, became the natural new frontier for ranching. Along with the early ranchers came the rustlers, stock thieves and all-around bad men. There was also the situation with the Indians, who were not welcoming of the ranchers but were very fond of their livestock. But the ranchers, tough "bad men" in their own right, drove most of the miscreants out of the Territory. Eventually, through government programs which put the Indians on reservations, the cattle industry was able to flourish throughout the Territory. The ranchers were able to expand into areas which were previously fraught with danger, and the industry grew rapidly. With the arrival in Arizona Territory of two transcontinental railroads in 1881, the livestock industry took off. The railroads opened the large eastern markets to the cattlemen of the Arizona Territory. Previously, most cattle were marketed locally or driven to a faraway shipping center; typically, a long and dangerous trip which always resulted in the loss of weight for the cattle and the loss of quite a few head along the way. Official reports, which were usually grossly undercounted, put the number of cattle in Arizona in 1881 at 78,000. By 1883, the reported number

This 1880s era photograph shows Edward Bowers' Ranch along the Agua Fria River. This was the former King Woolsey Ranch, now a ruin. Woolsey had a reputation as an Indian fighter and settled in the Dewey area in the early days when Indian "troubles" were a daily occurrence. Later, the Prescott & Eastern railroad would pass very near the Bowers' Ranch (Sharlot Hall Museum).

John Marr's Ranch in Lonesome Valley was photographed about 1893. Located on the Agua Fria River at the mouth of Lynx Creek, Marr's Ranch was well known for fine cattle and its hospitality. The stockade fence around the ranch complex is somewhat unusual and represents a significant investment in the Marr's Ranch cattle (Sharlot Hall Museum).

4. Ranchers and Farmers Follow the Dusty Trail to Mayer

Branding calves involved a lot of work—gathering the cattle, separating the calves from their bawling mothers, preparing the fire and the branding irons and then wrestling the calf to the ground and applying the brand. In this case, the calf looks to be a yearling who may have been missed in the previous year's branding operation. This photograph was taken in Yavapai County in 1903.

Sharlot Hall was a true "ranch woman" as was her mother, Adeline Hall. The Hall women did most of the work on Orchard ranch as James Hall did a little farming, a little ranching, raised a few horses and left most of the day to-day work to the women. Here, Sharlot is hand feeding baby pigs on the ranch (Sharlot Hall Museum).

had grown to 280,000. By 1920, the number was up to 1,170,000. Numbers dropped significantly during the Great Depression when ranchers were required to drastically reduce their herds but picked up again after World War II. As to the actual start of the cattle industry in Yavapai County, scant records indicate that the first few cattle were brought to Williamson Valley northwest of Prescott in 1864. Then, in 1866, James Baker reportedly brought about 400 head of Mexican cattle from Texas. By 1885, an article in the *Courier* estimated the number of cattle in the county at 80,000 to 100,000. This number grew in the next ten years but was reduced to nearly zero in 1896 after a severe drought caused starvation and death of an estimated 100,000 cattle, and 40,000 more were shipped out, leaving scattered remnants of stock from which to rebuild the herds.

This ranch scene, typical of the ranches in the lower elevations of Yavapai County, is Orme Ranch, a working ranch but also the location of Orme School. This photograph was taken in 1967 by Matt Culley (Sharlot Hall Museum).

Orme School students worked on the ranch on a daily basis. Many had their own horses, which they brought with them, while others used the ranch horses. These students were out on the ranch on a ride in the 1970s. The school is in the background.

An article in the June 1914 issue of *Yavapai* magazine reported that "over two thousand head of cattle have been shipped from Mayer this season. The principal shippers have been Charles H. Hooker, Hooker and Kellogg, Dugas, Burmister, Stewart, Fisher and one or two others. The cattlemen are getting good prices for their stock and with an assurance of good range conditions the future would be very rosy indeed." Another article in the same issue stated, "Fain & Heath of Mayer recently sold 650 head of cattle.... The animals were 1's, 2's and 3's, steers. The average price was $40."

A ranch woman on the Western frontier led, and may still lead, a life of hard work and isolation. In an article in the April 2007 issue of *Arizona Highways* magazine, Dave Eskes wrote of the ranching life and some of Arizona's ranch women including Rittie McNary Cameron, Elledean Bittner, Angie Brown, Clair Champie Cordes, Nellie Moore and Alicia Quesada. Eskes wrote that "ranch wives toiled from dawn to dark with few breaks and no conveniences. They crafted clothes, quilts and diapers out of feed sacks and scrubbed them clean on washboards. They cooked rib-sticking meals from scratch on wood-burning stoves and mended by the light of kerosene lamps. They canned (or dried) vegetables from the garden and gave birth in tin-roofed shacks without electricity, indoor plumbing—or doctors." As Elladean Bittner of Peeples Valley put it, "Arizona was hard on horses and women."

An overview of Stewart's T Anchor Ranch taken in 1941 shows the open grazing land at the foot of the Bradshaw Mountains and the isolation of the ranches in the area. Although roads have improved since the 1940s, many of the area ranches are still many miles from a main, paved road. Photograph by Rosena Promberger Minucci.

Some of the early ranching families in the Mayer area included the Browns on the Lower Agua Fria River; the Cordes family at Cordes, which was originally Antelope Station; the Dinsmores; the Quarter Circle V Bar, started in the 1870s by a man named Monroe but long owned by the Orme family who runs the famous Orme School at the ranch; the T Anchor, to the east of the Quarter Circle V Bar, originally started in the 19th century by the McDonald's but in the hands of the Stewart family by 1907; the Double T, owned by Jim Alexander prior to 1907 and subsequently in the hands of the Stewart family; the Lazy UL originally owned by Willie Rosenberger but subsequently also owned by the Stewart family; William Perry's on Perry Mesa who started with sheep in the 1870s, switched to cattle in the 1880s and ran the A Dot brand and, later, the Bar Box brand; Dugas Ranch, owned by the Dugas family and run by Fred Dugas; and Bensch Ranch owned by the Bensch family. The Goswicks, who were famous mountain lion hunters, ranched just northwest of Mayer. In Lonesome Valley, the Orchard Ranch is no longer a ranch and the old ranch house has been torn down. The old Double Wrench outfit established by freighter Tom Sanders in 1878 at the mouth of Yeager Canyon in Lonesome Valley is now owned by the Fain family who uses the Rafter 11 brand. Some of these ranches are still in the ownership of families who have ranched on the same land for 100 years or more. As late as 1966, the Goswicks, Dandreas, Benschs and the Guests were shipping cattle from the Mayer stockyards.

Most of the old-time ranches have changed hands numerous times down through the years, and the ownership is difficult to track. But the success of the ranchers, whether they held out for decades or moved on to other ranches or other professions, was dependent on the freighters and the merchants. As Matt Culley wrote in an April 1964 article in *Arizona Highways* magazine entitled "Old Yavapai is Cattle Country":

> While the pioneer cattlemen were, of course, the prime force in developing the range cattle industry in Yavapai County, there were other factors that undoubtedly played an important part, though perhaps indirect part. Perhaps the most important of these was the old-time general merchant who not only supplied the essential needs of the cattlemen in the way of food, clothing, hardware, cooking utensils and the like, but did it on a credit basis getting paid once a year when the stockman sold his steers (they sold steers in those days instead of calves which is the general practice today). Involved was a sense of mutual trust between the cattlemen and the merchant—no such thing as signing a sales slip was ever thought of—the merchant merely kept track of what the cowman bought and the latter paid for it without any question.

Once the railroad arrived in Mayer in 1898, it became a major shipping point for cattle, sheep, wool and mohair. Shipping livestock gave the ranchers an opportunity to get together and compare experiences in the last year and visit. Although originally located elsewhere, eventually the shipping pens were moved into Mayer adjacent to the railroad siding. In an undated article, Dr. Ernest Bensch, who grew up on Yavapai County ranches, including the Bensch Ranch along the Black Canyon Highway southeast of Mayer, wrote about the cattle ranches and the stockyards at Mayer. Dr. Bensch stated:

> The stockyards in Mayer were not always in Mayer; but were about a one hours ride south of Mayer depending on your horse. They were located along a lonely [sic] rocky

The Bill Thompson Ranch near Mayer was photographed by Matt Culley in 1970. Bill Thompson was a television broadcaster from Phoenix who, like many prominent Phoenicians, owned a ranch "retreat" in Yavapai County (Sharlot Hall Museum).

hill high above the canyons of Cedar Canyon; just before the train plunged down between the confining walls of the canyon on its way to Crown King, to reappear with leaden ore.

The stockyards were a meeting place for buyers, cowboys and ranchers that may not have seen each other for a year. Sheepmen also came to unload their wool bearing animals, as a starting point for the long trek to the High Mountains.... This was the only shipping point for probably a hundred miles in any direction.... The cattle would be driven to the stockyards, and separated; calves and the animals to be shipped were chased up the ramp into cattle cars and the doors closed. The remaining cows were allowed to return to their range home to repeat the whole process next year. Small steam engines would back up to the cattle cars and "puff" down the long winding track.... The noisey [sic] lumbering cattle trucks came and the locomotive came no more along the winding rattling tracks. The stockyards were moved away, and the grass grew over the road, the ranches changed hands, the cowboys rode tame horses the large herds disappeared and a part of the Old West was gone forever.

Cattle ranching wasn't the only livestock industry in the Mayer area. Sheep, goats, horses and mules rounded out the livestock activity in the area. By 1900, the sheep industry was well established in Yavapai County. The March 5, 1902, issue of the *Prescott Weekly Journal-Miner* reported that well-known sheep men Gus Reimer and C. C. Hutchinson were in Mayer. The booklet published by the Yavapai Commercial

This photograph shows the Mayer Depot about 1910 awash in Angora goats and illustrates the importance of the railroad for transporting livestock. The goats are milling around the depot and along the P & E mainline while awaiting shearing or transport to various area ranches or shipment to parts unknown—perhaps to shearing barns in the Phoenix area or to meat processing plants in the Midwest (Sharlot Hall Museum).

Club in 1907 states: "This section offers excellent opportunities to those who are interested in the stock business. The introduction of improved grades and more strict attention to the matter of breeding would be followed by correspondingly profitable results." The booklet states that the conditions in Yavapai County are ideal for horses and mules and that sheep raising is "perhaps the most important branch of livestock raising in Yavapai County." Along with sheep raising came sheep shearing and sheep dipping. The March 5, 1902, issue of the *Prescott Weekly Journal-Miner* further

The shipping pens in Mayer, which the local residents complained about due to the odors and the mess, not to mention the attacks on their gardens by escapees, were just west of the P & E tracks and south of the depot near the P & E warehouses. This photograph was taken in Mayer in about the late 1930s.

Sheep were being sheared in the spring around 1910 in a sheep shearing shed. It is back-breaking work for the shearers and traumatic for the sheep. The wool is piled into a hand cart which runs on wooden rails down the center of the shed. Sheep shearing contests were common in the shearing season to see how many sheep a particular shearer could shear in a certain amount of time without injuries to the sheep. Gasoline powered shearing machines speeded up the process in the 1910s. Many of the ranches did not have electricity until the mid–20th century.

reported that there was a steam sheep shearing machine at Bower's Ranch in Humboldt which seemed to be doing a "flourishing" business. Once the sheep were sheared, they had to be dipped and inspectors were on hand to be sure that the sheep were healthy as they went through the dipping pens. The March 1918 issue of *Yavapai* magazine remarked that more than 300,000 sheep passed through Mayer twice each year on the "sheep trail."

The Cordes family in Cordes were on the sheep trail, and, starting about 1900, became the supply center and shearing and dipping station for the sheep drives. During the drive periods, approximately 75 to 100 men would need a place to sleep and food to eat. The Cordes family put them up or let them camp out at Cordes and fed them in shifts in the family dining room. In an article in the Winter 1985 issue of the *Journal of Arizona History* entitled "Cordes and Cordes Junction," authors Robert B. Bechtel and Mynne Cordes Jarman described the sheep "year" at Cordes:

> The "sheep" year would begin with the drive south, lasting from September to November. It was then that the shepherds would stock up on supplies for the winter range. The drive north would begin in February, and shearing and dipping would take place throughout March and April.

Sheep had to be "dipped" after shearing before they could be turned out for grazing on open land. Don Francis (right front) of the Campbell and Francis Sheep Company kept an eye on the dipping process at Cordes. The dipping process at this station at Joe's Spring is set up as an assembly line (Sharlot Hall Museum).

A shearing pen was built for shearing wool by hand. With the old hand clippers one man was able to shear twenty to twenty-five sheep per day. In 1905 a man named Wynne brought gasoline powered shearing machines and the number sheared increased to sixty per day. As shearers trimmed the fleece, it was kicked aside so they could move to the next sheep. The wool-tiers, standing nearby, gathered each fleece into a bundle and tied it with a jute string. The tied fleeces were then thrown into a bin on wheels which was taken to a nearby circular rack from which the four-foot by eight-foot sacks were hung. The wool "stompers" would then stamp down the fleeces until the bag was full and sew it shut. The bags, weighing over 200 pounds, were then stacked until they could be hauled to a railroad siding. A spur was run from the Prescott and Eastern Railway, and for a time Cordes had its own railroad siding just three miles away.

Indeed, sheep raising was a major industry up until the time the industry peaked just after the passing of the Taylor Grazing Act in 1934. Sheep ranching in Arizona was primarily based in the Salt River Valley and in the Flagstaff area. Sheep were wintered in the lower elevations of the Salt River Valley and summered in the high country of Flagstaff and the Mogollon Rim. "Range Wars" between the cattlemen and the sheepmen are a thing of the past, but they were a serious situation in the 19th century in Arizona, resulting in the deaths of numerous ranchers and their cowboys and herders. Originally, most of the sheep herders in Arizona were from Mexico. However, Basque sheepherders from Spain became predominant in the sheep raising business by the early 1900s. Predators were also a problem, as they continue to be today. Formerly, trapping and poisoning were used to help control predators such as mountain

As the sheared wool was gathered, it was "stomped" into large burlap bags. It took three men, one pulling and two pushing it up a ramp, to load a 200 pound bag of wool into a Santa Fe, Prescott and Phoenix boxcar at Peoria, Arizona, about 1905 (Sharlot Hall Museum).

These men were having a good time at Mayer showing off their Angora goats for the photographer about 1909. The first goats were brought into Central Arizona in the 1880s, where the climate suited them well. By 1910, there were more than 15,000 goats in Northern Arizona. In the state, Angora goats were raised almost exclusively for mohair.

The cattle on the Yolo Ranch in Yavapai County were gathered by Gene Smith and his cowboys on a stormy day in 1950 (Sharlot Hall Museum).

lions, coyotes and bobcats, but that is no longer viable. Today, the sheep industry is still active. Cordes is still on the sheep trail, and, although most owners ship their sheep to summer and winter pasture by truck today, some years the sheep still come through in the spring on the sheep drive.

According to Mona Lange McCroskey's article in the November 2008 issue of *Territorial Times*, entitled "The Most Efficient Fiber Producers on Earth: Angora Goat Ranching in Yavapai County, Arizona, 1880–1945," "Between World War I and the end of World War II, the biggest industry in Yavapai County was the raising of Angora goats." The goats did very well on the open grasslands and dry, brushy terrain of Yavapai County. The United State Department of Agriculture claimed Angora goats were "a robust, elm-peeling, can eating, neglectable [sic] animal." However, although they were hardy and disease free for the most part, they were not able to range freely and needed close supervision. Herded by Basque or Mexican shepherds and usually a couple of dogs, the shepherds had to be on the watch for wandering goats and for predators. In the 1930s, a herder's wage was about $30 a month. It was a lonely life.

Sheared for their mohair and also used as a meat source (chevon), Angora goats were particularly popular in Yavapai County from the 1880s to the end of World War II. In her excellent article Mona Lange McCroskey wrote about the early development of the goat industry in the area:

> Most Yavapai County goat raisers came from the hill country of Texas, some bringing goats with them. Established cattlemen turned to goat ranching, or added goats to their cattle operations for three reasons: Arizona cattle production was beginning to surpass local market demands; buyers were demanding a better grade of beef than was produced by the longhorns that were driven into the state; and the drought of the 1890s had been disastrous to the cattle business. As one rancher put it, "Mohair was the meat and potatoes of the ranchers; cattle were the luxury."

In her article, Mona Lange McCroskey quoted an article written for *Arizona Highways* in May 1940 by Mrs. W. B. "Hattie" Young, who was active in the Arizona Mohair growers. Mrs. Young wrote:

> What sort of creature is this goat with his sudden snorts of distaste, his insatiable curiosity, this animal which cans the sunshine, wraps [it] into the long staple of his Mohair and holds it safe for ages, this animal with its long, curly, white coat of hair, this creature which is so fastidious and yet such a roughneck.
> If he is being herded and you remain perfectly still, curiosity will get the best of him and back he will come to investigate. If you'll continue to remain still he will be nibbling your clothes in a few minutes but he will not allow you to touch him.... He is happiest when playing on large rocks, bending trees or your automobile.

The introduction of synthetic fibers in the 1940s took a major toll on the Mohair market. It was the end of an era and an industry. Today, the United States Forest Service uses domestic goats to clear grasses and underbrush on forest lands as a part of their fire suppression programs.

Those who are still ranching today are, for the most part, living a life very similar to their 19th century counterparts, but with many modern conveniences such as gas and electricity, indoor plumbing, refrigerators, computers, cell phones, four-wheel

A tributary of Big Bug Creek ran through the "wide open spaces" of Stewart's T Anchor Ranch in the spring of 1941. Photograph by Rosena Promberger Minucci.

drive trucks and ATVs. But they still rely on the horse for much of the transportation and work on the ranch and their working dogs to help them with the round-ups, whether the livestock is cattle or sheep. They are still independent, self-sufficient and isolated, and, as their forefathers did, they crave the "wide open spaces" of the American West. Also, as their forefathers did, they have a great respect for the land and the life of the rancher. Kathy McCraine, a ranch wife herself, writes in the introduction to her book, *Cow Country Cooking*, "Some say ours is a 'vanishing way of life.' I think it is not so much a vanishing way of life as a changing way of life with an unchanging core.... But as long as ranchers can still make an honest living out of caring for their land and livestock, as long as they can put food on the American table, there will always be cowboys." As Noel Caniglia, who with her husband Tommy, works the U Cross Ranch six miles north of Mayer, says about the U Cross in *Cow Country Cooking*, "O. K., this is it. I can stay here. It's a real healing place with a lot of good energy. That sounds pretty funky, but this old house is 100 years old and it has a lot more history than we do. We're just visitors here."

5

School Bells Ring

The First Territorial Legislature met in Prescott in the fall of 1864. Governor John N. Goodwin asked for the establishment of a public school system, saying that "self-government and education are inseparable." Goodwin encouraged the establishment of public schooling, even if on a small scale, by proposing the setting aside of a portion of the funds which would be raised by taxation exclusively for schools. The legislators did, indeed, make a small beginning, setting aside $250 for the mission school at San Xavier del Bac in Tucson. But they also made provisions for $250 in matching funds for Prescott, LaPaz and Mohave and $500 for Tucson. The only school known to have actually received funds was the school at San Xavier del Bac, although Prescott, in Yavapai County, may have used some of the money. Thus, the modest beginning of the public school system of Arizona Territory was underway.

Samuel "Charming Dale" Rogers built Prescott's first schoolhouse in 1872. Constructed next to a large cottonwood tree, which is still there today, on what are now the grounds of Mile High Middle School, it was used as a school until 1876. On September 18, 1948, it was completely destroyed in an early-morning fire (Sharlot Hall Museum).

Mayer's first school, identified as "Big Bug School," was built by Joe Mayer. Three of the Mayer children are in the photograph—Mamie, in front of the door in the dark dress, Martie (maybe to Mamie's right, with Mamie's arm on her shoulder) and Burr (far right). The teacher is Mr. Johnson. According to Winifred Mayer Thorpe, the school was located "one mile from town," probably "up" Big Bug Creek to the north (Sharlot Hall Museum).

According to Winifred Mayer Thorpe, her father provided a light wagon "to take them [Mayer School students] to and from school. It was driven by a young man from Sandy Mush, North Carolina who was head man at the livery stable. Father never charged for this service." In this photograph, an Army ambulance transports children to school from Ft. Whipple in Prescott (Sharlot Hall Museum).

Public schools were slow to develop, but private and church sponsored schools sprang up around the Territory. The first Prescott school was begun in 1865. Also in Prescott, the *Weekly Miner* mentioned a number of people who held school during the late 1860s, including Samuel C. "Charming Dale" Rogers, who held school only when he was not engaged in mining business with his mining claims. Then school was out for a few days or weeks. However, his school term lasted all year round. Rogers was a well-educated man with a number of books to his name. His curriculum included the usual reading, writing and arithmetic, but also the appreciation of poetry, literature and the beauty of the universe. In 1870, he wrote persuasively about the problems he encountered with the school system and made recommendations which were later incorporated into the Territory's public school legislation. In 1872 he built Prescott's first schoolhouse under a cottonwood tree next to Granite Creek.

School terms were often short and sporadic—three months or maybe five months to a term and perhaps a total of seven months throughout the year. School buildings were often impromptu in nature—a "recycled" saloon building, store, or courthouse, sometimes including a jail, a room in someone's home, or the tack room on a ranch. Some communities built school houses with donated materials and labor, one room board and batt or adobe buildings with a dirt floor. In Mayer, the first schoolhouse was identified as being at "Big Bug." The "sponsor" of the school was Joe Mayer, who built it and provided transportation and room and board for some of the students so that there were enough children to "make up a school." The students included three of his own—Mamie, Martie and Burr.

The Mayer School district (later designated as Yavapai County District No. 43) was officially organized on October 1, 1888. Joe Mayer was the Clerk of the school board.

On the last day of school in 1896, Miss Mamie Mayer (age 17) wrote about the closing exercises for the school term. The article was published in the *Arizona Republican*:

> The last day of the Mayer School, April 24, dawned bright and clear. This day had come as all days must come, bringing with it the end of the pleasant term of seven months, which has been taught by one of Arizona's most accomplished teachers, Miss Mabel Meany. The morning passed as it seemed all too quickly. In the afternoon an exhibition was given before quite an audience of the trustees and families of the district, who were more than pleased with the progress their children had made under the careful attention of the teacher. The entertainment passed very pleasantly and after the exercises by the pupils were finished, Miss Meany gave a farewell address which will long be remembered by her pupils.

The program was as follows:

"Oh, We Are Merry Mountaineers"	Song by School
Recitations	
"Battle of Life"	Martie Mayer
"Three Children Whom I Knew"	George McCoy
"Babie Bell"	Jennie McCarty
"Lily's Ball"	Jeffie St. Germaine
"Children"	Concert Recitation by First Reading Class
"Incantation"	Wilbur Mayer
"Rock of Ages"	Lena Hartsfield
"Sheridan's Ride"	Grace Perry
"Mother Earth's First Child"	Reading by George McCoy
National Hymn	Song by School

In 1870, school taxes became mandatory, but the schools were perpetually short on funds. In ranching country, stray calves were rounded up for the "school herd." Creative ways to raise money for schools included raffles, dances, parties, carnivals, "cake walks" and cash donations. Families contributed land, materials, labor and room and board. Even as late as 1914, the townsfolk of Mayer held a "basket social" to raise funds to equip their new school building and raised $85.55. One basket brought in $11.00, a pricey sum at the time.

In 1869, Anson P. K. Safford, affectionately known as the "Little Governor," was appointed Governor of Arizona Territory by President Ulysses S. Grant. Safford, who had been in the West since 1850 and had long been involved in and interested in educational matters, made education one of his top priorities. Known as "the father of our public schools," Safford felt that the fact that there was not one public school in the Territory was both humiliating and mortifying. Safford subsequently wrote a public school bill, dubbed the Safford-Ochoa Act. The main purpose of his bill was to put some impetus behind a movement to found free, public schools, which Safford felt was the Territory's greatest need. His proposed system included a mandatory tax to support one or more schools in each County for a six month term each year. This tax of 10 cents on each $100 of property, to be collected at the same time as other taxes, would recognize that schools were one of the necessities of modern government. He proposed a Territorial School System which would require attendance of every child of sound mind and proper age. His proposed system was intended to aid and coordinate the county and local schools in the Territory but not to dominate them. Safford also believed that the most qualified person to put this system into effect was Safford himself, and his bill provided that he should do the job of the Superintendent of Public Instruction ex-officio as long as he was governor. As a part of his duties, he was to travel the Territory, visit districts and make speeches with the goal of promoting the establishment of schools throughout the Territory. In order to accomplish these tasks, Safford had included $500 annually to cover his expenses. On the very last day of the Territorial Legislative session, on February 17, 1871, the Safford Act was passed, but only after the "legislative hatchet-men" had cut out the funding. However, the $500 expense account had escaped the funding cut. As stated in *Dust In Our Desks*, in an article by Ernest J. Hopkins and Alfred Thomas, Jr. about Anson P. K. Safford, "Without loss of a day this 140 pound bundle of courage threw bacon, coffee and blankets into his buckboard, hitched up his two mules, and set out on the desert roads, alone to visit the widely scattered settlements of Arizona Territory at the height of the Apache Wars. He was going over the heads of the legislators, to talk to the people directly." Hopkins and Thomas, Jr., further state that "today's Arizona school set-up can be traced directly back to the Safford Public School Act of 1871 without a break."

When the Seventh Legislature convened in 1873, they had heard from the "people directly" and they took a much different attitude about schools in the Territory. The previously "axed" county school tax was reinstated and the Territorial tax was increased from 10 cents to 25 cents. The allocation of school tax money was a sticky problem as the distribution was based on average daily attendance. With the sporadic school terms, isolation of some schools and families, Indian issues, weather and other problems inherent on the Western Frontier, the scheme favored well established schools in the larger, more settled communities. The new plan was based on the population in each school district of school-age children. In order to implement the plan, a school census was necessary. In 1875, a controversy arose in Tucson over the passage of a bill by the Territorial Legislature providing $300 to a Catholic school. The quarrel between the Catholics and non–Catholics rippled through the Territory and resulted in the passage of laws prohibiting the public funds from being used to support religious schools. The first school census was completed in May 1876.

In *Dust in our Desks,* a quote from a diary left by a teacher near Prescott sometime before 1879 tells about the "dropping in" of the school superintendent. She wrote:

> We discovered that our little log school building was infested with a large woodtick, so we moved our school out under some scrub oak bushes, but as the windy season came on we had to find better quarters. There was in the school yard a large hole where someone had started to dig a well. It was fifteen feet deep and had so caved in that it was about the same width. One of the boys dug some rude steps, and in this impromptu school house we were finishing the term. The County superintendent, coming to visit our school, heard voices but saw no one, and as he was about to fall in on our school, I bade him come down. After holding a little examination, he christened this the banner school of the county.

Children are seated on what appear to be bear rugs in front of the Mayer School (the same building identified as the "Big Bug School"). Their paper banner, probably made by the teacher, who stands on the left, says, "Public School, Mayer, Arizona Education is...." The children appear to be dressed in their "Sunday Best" for their photograph, which was probably taken in the 1890s (Sharlot Hall Museum).

Anson P. K. Safford continued his one-man public education trips around the territory until the end of his term as governor in 1877. He had visited every settlement in the Territory of Arizona's 113,000 square miles. In between trips, he executed his regular duties as governor. His report to the Bureau of Education was modest in its understatement of his accomplishments. He stated,: "As soon as the Legislature adjourned, every part of the state was visited, and appeals to aid in the establishment of schools under the law ... were everywhere made. A desire for schools soon began to appear among the people." Safford was very proud to report that of the 2,995 children reported in the school census of 1876, 1,450 of them had learned to read and write by the end of Safford's last term as governor in 1877. When Safford completed his last term as governor, there were 28 public schools, 37 certified teachers and 3,089 students. It was quite an accomplishment for the "Little Governor."

Meanwhile, in Prescott, Safford had convinced Moses H. Sherman to come there to teach in 1873. Prescottonians were interested in education from the start, the first Prescott school having started just one year after the establishment of the town in

Prescott Free Academy in Prescott was built in 1876 on East Gurley Street. It was the Territory's first graded school. The building cost nearly $12,000 but served two purposes—four classrooms on the ground floor and Territorial offices upstairs. The Academy was torn down shortly after Washington School was completed in 1905. Photograph by Bate Studio (Sharlot Hall Museum).

5. School Bells Ring

1864. The community leaders helped Sherman establish the first graded school in Arizona Territory. In 1876, the voters of Prescott approved bonds to build a two-story brick school (Prescott Free Academy) at a cost of nearly $12,000, which was the best-equipped school in the Territory and the pride of Prescott. Several teachers taught classes under Principal Moses H. Sherman. Governor John C. Frémont had his territorial offices on the top floor and on Fridays his charismatic wife, Jessie Benton Fremont, conducted her "Arizona Class" for the school, sharing the

Joe Mayer built a new school in Mayer in 1902. It was originally located a few blocks from the Mayer Station but was later moved to a low hill southeast of town. After the Red Brick Schoolhouse was built, it was known as the "Ladies' Aid Building" but was still used for classrooms. Photograph by Rosena Promberger Minucci, 1941.

"Old Main," built in 1885, was the "main" building in the first years of Tempe Normal School (now Arizona State University). It was the very first state college building built in the Territory. This photograph is circa 1910 (Sharlot Hall Museum).

experiences of her European travels and her life growing up in Washington, D.C. Sherman went on to become the first full-time Superintendent of Public Instruction in 1879 under Governor John Charles Fremont.

In the 1880s, considerable progress was made toward the public school systems in the Territory. Teacher training institutes were established, women were given the right to vote for school trustees, and the number of taxpayers needed to establish a school district was lowered to five. The Tempe Normal School was established by the Legislature in 1885. Afraid that the Legislature would change its mind, Tempeans rushed to build an impressive building, and classes at Tempe Normal School started in February of 1886. Classes at the University in Tucson began in 1891, even though the founding of the university had been approved in 1885. The Northern Territorial Normal School was established by the Legislature in 1893, and classes began in 1899. Also in 1899, Title XIX was approved, requiring that all schools teach in the English language.

In 1902, a new one-room school was constructed in Mayer to replace the old board and batt school of the 1880s. The new school was originally located near the Mayer Station but was soon moved to a hill above the town. A newspaper article published on March 1, 1902, praised Joe Mayer for his generosity, stating in part, "The particular philanthropic act which he has done, and which, of course, he could not keep a secret, was the building of a fine school house at Mayer, at his own expense." Later known as the Ladies' Aid Building, it has been used for many community purposes as well as a school and has been moved twice. Over the years, it housed first through eighth grades and, later, kindergarten through second grades. In 1903, the

A group of students march up the front steps of the Mayer School, known as the "Red Brick Schoolhouse." The school appears to be new, which would make the date about 1915. One of the "moon houses" can be seen to the right of the school. The Ladies' Aid Building would be just off the edge of the photograph to the right (Sharlot Hall Museum).

A close-up photograph, probably taken in the 1920s, of the Mayer School. In an undated newspaper article headlined "Little Red Schoolhouse in Cow Country" the reporter wrote, "[W]e've heard about little red country schoolhouses, but we never quite believed they existed until we came to this town. Here is one in the good old tradition. It is red brick, has seven rooms, five teachers and takes in all the classes from kindergarten to 12th grade."

school had an enrollment of 23 pupils. An article in the *Weekly Journal-Miner* of June 9, 1903, stated that " the school house is one of the best in the County. Mr. Mayer put in a new piano in the building recently. Mrs. J. M. Collins, ... the teacher, is one of the most competent teachers in the territory , and the children are progressing rapidly under her instruction. The calisthenics and wand drill executed by her pupils is a graceful and pleasing spectacle."

In an interview for the *Prescott Courier* in 1982, Irene Thompson McDonald mentioned that she went to first and second grades in "a big brick building this side of the White House Hotel. In the third grade I went to school in the little Ladies' Aid Building, then started school in the brick building that's there now." She would have started school about 1911. No other references to the "big brick building this side of the White House Hotel" have been found.

In Yavapai County, five school districts had successful bond elections in 1914 for new school houses: Prescott, $95,000; Jerome, $25,000; Mayer, $10,000; Ash Fork, $5,000; and Cornville, $1,600. Prior to his death, Joe Mayer donated a site near the 1902 school for a new school to be built in the future. The bond election of 1914 made that possible. In a July 15, 1915, article in *Yavapai* magazine E. I. Long, Clerk of the Board of Trustees of the Mayer School District wrote, "If one were to ask the question 'In what way has the Mayer community improved the most during the past two years?' the answer from everyone would be 'The improvement in her school.'" The school was designed by architect W. S. Elliott. Prescott contractor Joe Petit was awarded the contract in June 1914 for the construction of the school, which was to be of brick, two stories high, with four rooms at a cost of nearly $9,000. The school was completed by July of 1915. E. I. Long further wrote of the Mayer School District that "the new school house, the creditable showing made by the pupils of the Mayer School at the Northern Arizona Fair held in Prescott last fall, the Mayer School Literary Society, the introduction into the school of old fashioned spelling matches, entertaining public programs, a neat looking school ground, increase in the attendance of pupils and the addition of another teacher, can all be mentioned among the things which indicate the progress of the Mayer School." Known as "the red brick schoolhouse," the Mayer School originally went through the tenth grade. The mascot, the "wildcat" was chosen early on and originally the Mayer School yearbook was called *La Loma Del Cobre* (the hill of copper in Spanish) in honor of Mayer's mining industry. School colors were originally maroon and gold.

A September 15, 1916, article in *Yavapai* magazine about the population growth throughout Yavapai County, stated," Owing to the development of mining properties throughout Yavapai County, it will be necessary to employ fifteen additional teachers this year. Additional teachers were required this year at Mayer."

In 1932, Principal Joel A. Benedict was successful in getting accreditation for a four-year high school program for Mayer School. There were four teachers including the principal who were expected to teach full time for a nine-month term for $175 per month. Benedict was employed to teach all of the subjects in the ninth and tenth

There are approximately 90 students in this photograph of the Mayer School, plus four teachers and the principal, who may be Loren Pierce. The children have come from all around the area—Mayer, Cordes, Poland Junction, and the surrounding ranches, arriving from outlying areas in the "funny little red school bus driven by the town preacher" (Bill Promberger).

5. School Bells Ring

An undated photograph of the 6th, 7th and 8th grades at Mayer School, probably taken in the 1940s. Mayer School was used for classes until 1982. Restoration work on the building allowed it to reopen in 1987, however, currently, the building is not in use (Bill Promberger).

The front of the Mayer School with its Roman-style arch in 1941, the year Bill Promberger graduated High School at Mayer School. Photograph by Rosena Promberger Minucci.

grades—four subjects in each grade. At the time, Benedict did not know of any student from Mayer to have finished high school or gone on to college. Benedict wanted to change that statistic. Mayer was a mining town, but during the Depression, the mines were closed and the town had a hard time meeting the school budget. Expanding the budget for the school was, consequently, prohibitive, so Benedict came up with a scheme, called a "leap-frog" curriculum, to teach the ninth and eleventh grades one year and the tenth and twelfth grades the next year. The ninth graders were combined with the tenth graders and the eleventh graders with the twelfth graders, all in one room (it was a four-room school, and the lower grades were combined into two classrooms downstairs). It worked, and the first high school graduating class of the accredited Mayer High School of two students, Ellen Surrett and J. E. Garber, was held on May 19, 1933. In a 1990 article entitled "The Depression Start of Mayer High School or Four Years for the Price of Two," Benedict wrote that in teaching and administering educational programs for more than 50 years, he felt that the organization of the Mayer High School met the greatest community need.

Eventually Mayer School housed grades kindergarten through 12th in its four rooms. William (Bill) Promberger, who went through school in the Red Brick Schoolhouse and graduated from high school in the Class of 1941, recalled some of his memories of the school. He wrote:

> On the back side of Mayer School on the top floor was a door to the fire escape. The fire escape was a slide that went from the second floor to the road behind the school. The road was narrow so there was room for the slide. When the boys, with their waxed bread wrappers from their lunch, would make the slide slick, you would end up in the road when you went down. But on each side of the slide, next to the road, were "out houses" on one side for the boys and the other for the girls. In 1934 the W.P.A.* dug out the basement and installed flush toilets. It was something to remember when the "ripe" outhouses were replaced.

The 1960s brought lots of changes to the Mayer School District. After becoming a sleepy little town after World War II, Mayer was growing, and that growth meant that the Mayer School was just too small to accommodate the student body. Three new classrooms were built next to the red brick schoolhouse for the lower grades. The last high school class to graduate from the red brick schoolhouse was the Class of 1964. In 1965, a new school was built to the west of the Mayer School for the high school students. Today, there is a new Mayer High School in Spring Valley, built in 1982. The former high school, across the wash from the red brick school house, is now the elementary school. The red brick schoolhouse is no longer used as a school but has been restored to a certain extent and updated and modernized. It has been used as administrative offices and a community meeting center. There is a restored classroom and a historical display of photographs, memories and artifacts from the 87-year history of the school. The 1902 school building (the Ladies' Aid Building) is still standing and is now a private home.

Mayer School is not, however, the only long-term school in the Mayer area. The Quarter Circle V Bar Ranch School was started on the Orme family's ranch 12 miles

---

*Works Progress Administration

The 1902 Mayer School, later known as the Ladies' Aid Building, now a private home. Photograph by Nancy Burgess, 2011.

The modern Mayer Elementary School, built across the creek to the west of the Red Brick Schoolhouse, includes grades kindergarten through eighth. Photograph by Nancy Burgess, 2011.

QUARTER CIRCLE V BAR RANCH
SUMMER CAMP
MR. AND MRS. C. H. ORME
TELEPHONE: HUMBOLT 351
MAYER, ARIZONA

Letterhead from the late 1940s for the Orme School showing the Quarter Circle V Bar Brand over the gateway. The letterhead advertises the "Orme School Summer Camp." Girls who attended the Orme School in the 1940s got to wear jeans to school, when public schools typically required dresses or skirts for all female students.

from Mayer in 1929. The Orme family were pioneers in the Salt River Valley (Phoenix), farming large tracts of irrigable land starting in the 1870s. Charles Henry Orme, known as "Chick," was born in 1893. As a student at Stanford University, he met and married Minna Vrang. In 1929, the family purchased approximately 30,000 acres of what was known as the "Hooker Ranch" in Yavapai County. On June 8th, they arrived at the ranch with their three children, Charlie Orme, Jr. (age 12), Morton Vrang Orme (age 7), and Kathryn Minna Orme (age 3–1/2), known as "Katie." Minna Orme recorded what she saw as they arrived that day in her journal: "Upon entering the old ranch house, there sat the haying crew; their last day around the oval oak table [with a] huge granite coffee pot; and I wondered what I had gotten into. However, hay was still being pitched [by hand] and it took strong, rough men to do it."

The Quarter Circle V Bar Ranch was a working cattle ranch. However, a school was the first priority for the Ormes, who decided not to try to transport their children over the rough ranch roads to Mayer to attend school. Instead, the Ormes worked out an arrangement with the County School Superintendent to replace the then-abandoned Dugas Accommodation School with a school at their ranch. Charlie, Jr. and Minna were able to salvage the desks and blackboards from the Dugas Ranch school, and these were the original furnishings for the school. They were used for many years. Originally, the school at the Quarter Circle V Bar Ranch was known as the Hooker Accommodation School, after the original ranch owner, Charles Hooker. One year later it became known as the Orme Ranch School. The Ormes used buildings that were already on the ranch, and the one-room school took up the west side of the bunkhouse. There was one teacher and seven students. The classroom had 15 desks and blackboards which were tacked to the walls, as well as a pot-bellied stove which kept the classroom warm in the winter. Initially, students included the Orme children, children of the ranch hands and children from nearby ranches. The Ormes made arrangements for those students to board in if they needed to. According to an excerpt from a letter from teacher Gladys Loftin to Susan Adams Samuelson, included in Samuelson's article in the *Journal of Arizona History* entitled "The Orme School on the Quarter Circle V Bar Ranch," "The subjects adhered to the state course of study and included the basics: reading, arithmetic, geography, history, spelling, writing and English. Crafts [included] painting, salt maps, clay [river mud] objects, miniature whittling, and cloth batiks."

Gradually, the school transitioned from a county school to a private school. Known as Orme School, the school and the family struggled throughout the Great Depression in the 1930s. They raised turkeys and vegetables. They instilled the value of hard work in their students, and each had chores to do. Students came from a number of sources, including family, friends and more distant family members. Very early on in the enterprise, the Orme family encouraged children with allergies, particularly asthma, to come to the school and claimed Yavapai County to be the best spot in the United States for the relief of asthma in children. Enrollment remained small until after World War II, when Charlie Orme, Jr., took over the management of the school. By then, the curriculum included high school. The Orme School continued its long-time commitment to a strong scholastic program combined with ranching activities. In an article in the *Ford Times* magazine in January 1952, entitled "The Three R's on

An overview of the 40,000 acre Orme Ranch in 1967. Peter Niggeman, who graduated from Orme in 1966, wrote in The Building of a School, The History of Orme School*: "A week doesn't go by without me thinking of Orme School. Orme taught me about community and the Judeo-Christian heritage ... I was fortunate to have been at Orme when it was a working cattle ranch and school. We all had to work together to make the community function. Without each of our chores, the community couldn't function...." (Sharlot Hall Museum).*

the Range," author Raymond Carlson wrote that "equally important, however, are the benefits—tangible and intangible—of a school on a real, working ranch. It assures a congenial, home-like atmosphere, and the students take an active part in the ranch and farm life. Each one has his own daily chores, which helps to develop the qualities of good citizenship and individual responsibility."

Because of the location of the school, which now encompasses 55 acres set in the middle of the 40,000 acre Orme Ranch, the Orme School faculty, staff and students truly became a family. The school has grown, changed and improved greatly since the Orme family set up a school for the children on their ranch in 1929. The 1967–68 student handbook stated the school's position on the goals and values set by Chick and Minna Orme:

> Although we believe in the importance of individuality, we know that individuals cannot live in isolation. As a small but unusually diversified community, the Orme School provides a rare opportunity for boys and girls to understand and appreciate the demand of a community on its members.

Today, the Orme School, with students from all walks of life and from all over the world, is one the nation's most innovative, competitive and unusual boarding schools. Using the ranch as an outdoor laboratory, every possible traditional academic subject is taught in a beautiful southwest setting.

6

Joe Mayer, Entrepreneur Extraordinaire

The March 1918 issue of *Yavapai* magazine featured Joseph Mayer on the front page under the headline "Builders of Yavapai, Something About the Man Who Recognized the Big Possibilities In Store For Mayer." The article said of Joe Mayer: "It would be hard to find a man more thoroughly typical of all that was best of the 'Pioneer West' than the founder of Mayer. To every inhabitant of Yavapai County from the bewhiskered prospector, whom he grubstaked to sowbelly and beans, to the silk hated capitalist who listened with respect to his suggestions, he was known as 'Joe' Mayer. This was not because of any lack of respect. It merely indicated a friendliness and confidence that was begotten by the evident honesty and direct kindness of the man."

As Arizona State Historian Marshall Trimble wrote in *Diamond in the Rough, An Illustrated History of Arizona:*, "The men and women of the Old West are among the

This wonderful photograph of the Mayer Business Block must have been taken in 1904, shortly after the four phases of the building were completed, as there is very little signage. The Mayer Saloon is on the left, the restaurant is next, then Wicks and Mayer Mercantile in the original building built in 1902, and on the right, the barber shop. Joe Mayer is standing next to the pony cart on the far right with a small dog at his feet. A freight wagon and a trap are tied up out front. The hitching rails are still in front of the building. With about 25 people standing around in front of the Mayer Business Block, it must have been a busy day in town (Charlene S. Thorpe Fry).

A reservoir for hydraulic mining near Lynx Creek provided the water necessary for the water hoses. Erwin Baer photograph, 1890 (Sharlot Hall Museum).

most cherished figures in Americana—the rugged symbols of the making of a country, hardworking, self-reliant, with honest determination and independence. In short, all those virtues that people like to see in themselves, not just in this country but throughout the world." This description fits Joe Mayer to a "T."

Like many "men of the Old West," and many of his late 19th and early 20th century contemporaries, Joe Mayer was a man of many interests and many enterprises. He was a husband and a father, an entrepreneur, hotelier, restaurateur, mine investor, merchant, farmer, rancher, land developer, philanthropist and unabashed promoter of Mayer, Arizona. At the same time, he was unpretentious, generous and kind. He had wonderful dreams of the Mayer of the future and always envisioned that it would be a center of commerce in Yavapai County and would eventually be much larger and more important than Prescott. Unfortunately, that vision would not materialize for Joe Mayer.

The March 1918 issue of *Yavapai* magazine article said of him: "People called him lucky. Everything he touched seemed to prosper. When he joined in on a mining venture, it almost invariably turned out successfully. He realized more than others the wonderful richness of the country back of Mayer.... Even in the days of deepest depression nothing could shake the confidence of Joe Mayer in the ultimate importance of the [mining] district and the growth of the town of Mayer that was to be its metropolis."

Once Joe and Sadie Mayer settled in at Big Bug Station in 1882, Joe got to work to make the place better for his family and more attractive to his customers. He planted apple, cherry, plum, black walnut and apricot trees. The new station included his family home; rooms for guests; a general mercantile store with an adjacent large dining room, a small bar, and a kitchen; and a barn with a corral. Once Sadie became the Mayer postmistress in 1884, a small area was partitioned off in their home for the post

6. Joe Mayer, Entrepreneur Extraordinaire

This photograph by Erwin Baer of the Lynx Creek Reservoir after an 1891 flood shows damage similar to what would have happened to the dam on Big Bug Creek which destroyed the Mayer Station on March 1, 1891 (Sharlot Hall Museum).

A photograph of hydraulic mining on Lynx Creek about 1890 shows the destruction of the terrain that this type of mining could cause. Fortunately, hydraulic mining was not commonly used in Yavapai County—it took too much water (Sharlot Hall Museum).

office. There were no regular hours or days off, including Sundays. Sadie would open early in the morning and occasionally got up in the middle of the night to give someone who was passing through their mail.

The Mayer's home and the Mayer Station were destroyed by the flooding of Big Bug Creek on March 1, 1891. According to a footnote in the article "Joe Mayer and His Town" in the *Journal of Arizona History* by Winifred Mayer Thorpe (Winnie), "The Big Bug flood and dam break was one of four such catastrophes occurring in central Arizona during 1890–91. Hydraulic mining companies had purchased the best placer locations along Lynx Creek, Big Bug Creek, Humbug Creek and the middle Hassayampa River. Each built reservoirs with flimsy water-storage dams which burst during the February freshets." In the same article, Winnie wrote that her father had sent a man upstream on horseback to ask that the dam be opened and the water let through, but the request was refused. Winnie further wrote:, "The roar of the coming water could be heard long before it was sighted and everyone knew what was happening. Huge boulders, uprooted trees, bodies of cattle—all rushed by on the flood of black water. The station folded up like a cardboard box and was gone. The spot where it stood is now in the middle of the creek and few of the lovely old trees remain." Soon, Joe Mayer was rebuilding, using some of the materials from his buildings which were washed downstream. A newspaper article in the *Arizona Weekly Journal-Miner* of March 11, 1891, stated that "Joseph Mayer, with his characteristic energy and enthusiasm, has rebuilt a temporary station from the wreck of his former place and he has ordered material for permanent buildings." The new, permanent Mayer Station was completed late in 1891 and was the Mayer family home until July 1983, when it was destroyed by fire. In "Joe Mayer and His Town," Winnie described the new station, known locally for decades as the "stage stop," as "large and spacious with long porches front and rear. One big room served as a store, another as a dining room. Joe hired four Chinese boys to run the restaurant—one to cook, two to wait on tables, another as a dishwasher. A row of wood cottages were built across the street as guest rooms. Several years later they burned down and were replaced by the brick apartments."

In the meantime, in the mid-1880s, while Sadie was the postmistress of Mayer, Joe was delivering the mail to Stoddard at Copper Mountain on horseback. In "Joe Mayer and His Town," Winnie says that Joe and Sadie were great friends with the Isaac Taft Stoddard family who lived at Stoddard.

Eventually, as Mayer became the commercial center for the area, the custom of Mayer Station outgrew the "new" station and Joe Mayer embarked on the construction of the Mayer Hotel across the street from his current enterprises. The architect and builder was Hill C. Moore. An October 1, 1897, newspaper article states that "Mr. Moore has charge of the construction of a large hotel for Mr. Mayer, and says it will be one of the best appointed and handsomest buildings in Northern Arizona." The Mayer Hotel included a mercantile store, restaurant and meeting room downstairs and lodging rooms upstairs. It was completed in the fall of 1897*.

---

*Many sources give the date of construction of the Mayer Hotel as 1887, 1888 or 1889. However, an October 16, 1897, newspaper article clearly states that the building would "be completed in about a week."

6. Joe Mayer, Entrepreneur Extraordinaire

Most of the dams on the creeks and rivers in the Prescott—Mayer area were constructed by the mining companies with varying levels of skill and engineering. Construction of the wooden Walnut Grove dam is shown in this circa 1880 photograph. The Walnut Grove Dam disaster in 1890 killed an unknown number of people when the dam burst and water came rushing downstream, bringing trees, boulders and all kinds of debris with it (Sharlot Hall Museum).

This photograph circa 1900 shows the "new" Mayer Station, built in 1891 for the Mayers after their previous station and home were destroyed by flood. Clearly, the fence is a "make do" situation. It would later be replaced by a white picket fence. This building burned in 1983 (courtesy Arizona Historical Society/Tucson, no. 61056).

This unidentified woman riding a horse in the middle of the Black Canyon Road in Mayer has just passed the Mayer Hotel (in the background) and has paused in front of the cottages Joe Mayer built next to the hotel to accommodate the overflow from the hotel. The cottages burned in 1907. This snapshot was probably taken about 1900.

A wood frame building on a brick foundation, the Mayer Hotel originally had horizontal wood siding. With a footprint of 60 feet by 60 feet, it is two stories high and has a porch across the front, which was the orator's stand for political rallies or announcers at special events, and a railed balcony on the second floor. Later owners described the "lodging rooms" as 23 tiny rooms with one bathroom. One undated newspaper article states that the south (left) side of the building originally had a gallery. A traveling reporter for the *Arizona Republican* who rode the Black Canyon Stage to Mayer and stayed in the new hotel commented in the November 6, 1897, issue:

> A mile further on is Mayer stage station and the vast onyx fields. Here we dismount and take up quarters at Mayer's where you are made to feel at home. Mr. Joe Mayer, though a resident of this locality sixteen years, has not fallen into the rut. He is awake and the new store building and hotel is offered in support of the statement. The building covers ground space of 40 × 60 feet. The lower floor is occupied as a general merchandise store and saloon adjacent, a hall separating. On the upper floor, six sleeping rooms and a hall, the latter 14 × 40. A veranda was built in front and this is a pleasant place to take an airing.

In November 1897, the "Mayer Correspondent" to the *Journal-Miner* reported that "a fine two story building is the latest additions [sic] to Mr. Mayer's enterprises, where he has stacked a splendid general supply of goods needed in that section,

When in Mayer

STOP AT

The Mayer Hotel

M. T. KNAPP, Prop.

European Plan

Rooms Nicely Furnished

A 1917 advertisement for the Mayer Hotel shows Justice of the Peace M. T. Knapp as the proprietor. Joe Mayer completed his business block in 1904 and subsequently sold the Mayer Hotel.

presided over by W. E. Wicks. Adjoining the merchandise department is the bar room, with J. E. Rawdin, an excellent mixologist, in charge." The *Arizona Republican*, probably quoting the very same "Mayer Correspondent," commented in the January 8, 1898, issue: "The Journal-Miner man in his travels has enjoyed the hospitality of one of Yavapai County's prominent and successful citizens, known far and wide as Joe Mayer." On December 1, 1897, the "Sunset" telephone was moved to the new hotel by Edwin Treadwell. By 1913, the Mayer family had apparently sold the hotel, as the 1913 Yavapai County Directory listed the Mayer Hotel with notary and Justice of the Peace M. T. Knapp as the proprietor. The March 1918 issue of *Yavapai* magazine said of the Mayer Hotel: "The management of the hotel is under 'Judge' Knapp, who is also the proprietor. Many mining men of the district make the Mayer Hotel their headquarters." Today, the Mayer Hotel houses apartments.

Joe Mayer had a livery stable and had a road built to Crown King, 22 miles up the mountain southwest of Mayer. In the 1890s, this made Mayer the major supply station and "jumping off point" for the mining districts in the area, including Crown

Joe Mayer built a road to Crown King and the Crowned King Mine. Heavy freight wagons hauled the ore down the mountain to Mayer. This photograph is of Mayerites Joe Norwood and Pete Diskin hauling ore for the Crowned King Mine, dated circa 1916 (Sharlot Hall Museum).

This photograph shows the size of some of the pieces of onyx from the Mayer Onyx quarry which had to be transported by freight wagon prior to the arrival of the railroad.

King. Large freight wagons would stop at the livery station to rest their teams and often to spend the night. One of the commodities that the fourteen-hitch freight wagons carried was onyx from Joe Mayer's onyx mine which was shipped from Prescott by railroad. An article in the *Arizona Republican* dated December 16, 1896, reported that Joseph Mayer and Joseph W. Wilson had purchased interests in a number of placer onyx claims for $10,000. Another commodity for the freighters was dynamite for the mines. In "Joe Mayer and His Town," Winnie wrote that her "mother noticed a horse saddled and tied outside the gates [of the livery stable]. She asked John Livingstone, then in charge, why the horse was out there. He replied, to her surprise 'I keep the horse ready in case the powder goes off, so I can ride out of town in a hurry.'" Soon a mountain site about a mile out of town was selected by Joe Mayer, where he dug a cave into the mountainside and lined it with thick concrete to be used as a powder house. At the same time, Mayer was one of two stations on the Black Canyon Road between Phoenix and Prescott where horses could be changed by the three-times-a-week stage companies. An *Arizona Republican* newspaper article dated October 8, 1896, states that "Mayer is the center of quite a lively commerce. It is here that Bruce's teams make freight transfers from the 4-horse teams to the 10s and where they meet for transfers. It is not much of a strain to believe Chicago is here." In about January 1898, Joe Mayer established the Mayer Freighting Company after he and a partner, Fred Brecht, had purchased the Bruce Freighting "outfit." Mayer was a busy man and the town of Mayer was a busy place.

Big changes would soon be coming to Mayer. Joe Mayer was friends with Frank M. Murphy, the president of the Santa Fe, Prescott & Phoenix Railway, and their engineer William A. Drake. Murphy wanted the ore shipping business for the mining districts between Prescott and the Big Bug Mining District. Joe Mayer persuaded Murphy to bring his railroad the 26 miles from Prescott into Mayer and to make Mayer the major railroad shipping point for the area. The fact that Joe Mayer provided

This is a rare photograph of the first phase of Joe Mayer's Business Block, which was built in 1902 for the Mayer Mercantile. The restaurant and saloon would be built adjacent to this building to the left and the barber shop would be built to the right. A close look at the front of the completed Business Block clearly shows the different brick corbelling on the front of this building. The storefront remains pretty much unaltered today (Charlene S. Thorpe-Fry).

This typical turn-of-the century street scene has it all—men on horseback, a dog, children playing on the hitching rails, men standing out in front of the local saloon, chairs out on the sidewalk at the side of the building, etc. This photograph of the Mayer Business Block was probably taken about 1905–1906 after the building was completed in 1904 and after the trees had had a chance to grow. Since the trees are leafed out, and most of the men are wearing light coats, the photograph was probably taken in the spring or fall.

This interior photograph of the Wicks & Mayer Mercantile was probably taken about 1902, shortly after the building was completed. The store is beautifully fitted-out and everything is neat and organized. A very large Indian basket, probably Yavapai, is displayed front-and-center on top of a glass, free-standing display case. The store carried everything from fresh produce and groceries to mining supplies and advertised "a large and carefully selected stock constantly on hand." The upstairs railing provided a place to display animal pelts on the right and Indian rugs on the left.

Joe Mayer

DEALER IN

GENERAL MERCHANDISE

AND MINERS' SUPPLIES

Joe Mayer's letterhead from a June 22, 1905 letter in which Joe Mayer wrote, "I send you by express this day lion scalp killed by Geo. Hutchins, known as Old Hutch and enclose affidavit to that effect. Send bounty to me soon as convenient... Yours Truly, Joe Mayer."

WICKS & MAYER
MAYER, ARIZONA.
Dealers in
General Merchandise and Miners' Supplies

Your patronage solicited and careful attention given to all orders. A large and carefully selected stock constantly on hand. Goods packed and shipped to adjacent camps.

We Can Please You and Want Your Business.

A December 1904 advertisement from the Arizona Weekly Journal-Miner *for "Wicks & Mayer," "Dealers in General Merchandise and Miners' Supplies" offered "goods packed and shipped to adjacent camps."*

the right-of-way through Mayer with a legal provision that if the right-of-way were ever abandoned, the land would revert to the Mayer estates, probably helped seal the deal. Construction started on the Prescott & Eastern Railroad in March 1898 and it opened on October 15, 1898. As Winnie wrote in "Joe Mayer and His Town," the railroad "made business easier for the ranchers and easier for Joe Mayer to supply them." It also brought visitors from far away who were graciously hosted by the hospitable Mayers. And for Mayerites and the ranchers and miners in the area, the trip to Prescott on the P&E took only about three hours instead of from dawn to dusk by wagon.

One of the few early photographs of the Mayer State Bank shortly after its completion in 1917 also shows the construction of sidewalks in downtown Mayer at about the same time. A canine "customer" is waiting outside. Except for the addition of an awning, the building looks pretty much the same today.

By 1897, Joe Mayer had established a brickyard. W. S. Deeds was the brickmaker. The bricks were made of local clay soil and were fired locally, although no known archaeological remnants of a brick kiln in the area have been found. By 1902, when Joe Mayer started to build the Mayer Business Block across the street from the Mayer Hotel, Joe Mayer was building with brick rather than wood. The bricks used for the Mayer Business Block and the later Mayer Apartments have held up fairly well and they do not appear to be commercially made "imported" bricks which would have been ordered from a Phoenix or Los Angeles brick company and delivered by railroad. It can be assumed that the bricks for these buildings were made at the Mayer brickyard. The brickyard was still in business in 1908 but exactly where the brickyard was located or when the brick business was established by Joe Mayer and how long it lasted is unknown as no records could be located.

Joe Mayer built two small, brick apartment buildings across the street from the Mayer Station in 1907 to replace the wooden cottages that burned that year. Each building had three apartments. Rosena Promberger Minucci took this photograph in 1941 of one of the buildings. The outhouses are in the back yard. A little girl is standing next to laundry which has been hung out on the line.

6. Joe Mayer, Entrepreneur Extraordinaire

The 1861 Brunswick Bar in the Mayer Saloon, which, according to Winifred Mayer Thorpe, Joe Mayer bought and had shipped by freight wagon from Virginia City, Nevada in 1900. The question is, was it originally installed in the Mayer Hotel and then moved across the street to the Mayer Saloon, or was it actually purchased later for the Mayer Saloon, which wasn't completed until 1904? In the 1950s, Lonnie Wright ran the Mayer Cigar Store and Bar with "mixed drinks, wine, beer and café." The bar remained in the saloon until 1965 when it was sold to Pioneer, Arizona a living history museum, where it is on display in their restaurant.

COME TO W. J. MAYERS' STORE
STAPLE AND FANCY GOODS
Come here to get your winter supplies
MAYER, ARIZONA

By 1922, the mercantile in the Business Block was known as W. J. Mayer's and was run by Wilbur J. "Burr" Mayer. A 1922 advertisement for the store from the Big Bug Breeze *offers "Staple and Fancy Goods."*

The Mayer Business Block, also known as "the Joe Mayer Block," was built in four phases. It is actually four brick buildings built adjacent to each other. It was completed in 1904. These buildings housed Joe Mayer's enterprises: the Mayer Saloon and Cigar Store, a restaurant, the Mayer Mercantile and the Mayer Barber Shop. The Mayer Saloon boasted a beautiful, hand-carved Brunswick bar with matching cabinets, which Joe Mayer reportedly bought in Virginia City, Nevada in 1900. He had the bar shipped to Mayer by freight wagon. In 1965, the bar was moved again, this time to Pioneer, Arizona, where it is on display.

Also, in December of 1902, the main street of Mayer was graded and some concrete sidewalks were installed. The Mayer Mercantile became Wicks & Mayer around 1904, then later Mayer & Mayer and W. J. Mayer General Mercantile. The Mayer Hotel, the Business Block and the Mayer Owl made the corner of what is now Oak Street and Central Avenue the "downtown" of Mayer. Later, in 1917, the Mayer State Bank would be added to this corner. At about the same time, the "new" Mayer Post Office was constructed just to the north of the Mayer Business Block. Today, the Mayer Business Block is listed in the National Register of Historic Places.

The Mayer Apartments, built in 1907, were intended to replace the wooden guest cottages Joe Mayer built after the flood in 1891 and which burned in the summer of 1907 (report in the *Arizona Miner*, June 3, 1907). They provided additional lodging for businessmen and travelers who were staying in Mayer for a few days or longer. However, local legend says that they eventually evolved into Mayer's "red light" district. Today, they are still apartments and are listed in the National Register of Historic Places.

In an article in October 8, 1908, issue of the Prescott *Courier*, a reporter wrote:

> At Mayer, all is bustle.... On inquiry as to what a long, low building south of the depot was, I was informed that it was a warehouse. "Who owns it?" I asked. "Joe Mayer" was the reply. I expected he owned the hotel where I stopped as I met Joe there so asked no questions concerning it. A fine brick store across the way caught my eye and I asked if that was a B-B store. My companion smiled at my verdancy [sic] and replied, "Not much; that is Joe Mayer's." On my way to the new Rigby works Wednesday morning I passed six newly completed buildings, and on inquiry as to who was the owner, was informed "Joe Mayer." I espied a brickyard with a full crew turning out 10,000 first class bricks daily, and with a 200,000 kiln of brick just burned and asked who it belonged to and was told "Joe Mayer." So I concluded that Joe Mayer was "it" here and no one seems to begrudge him the honor for he sure took advantage of his opportunities and is now reaping his reward.

The Mayer Business Block, Mayer Apartments and the site of the 1891 Mayer Station, known as the "stage stop," were sold out of the Mayer family after Winifred Mayer Thorpe's death in 1983.

Joe Mayer was also partly responsible for bringing water to Mayer from Grapevine Springs, about eight miles northwest of Mayer. The pipeline was built by the Phelps-Dodge Company, largely through Joe Mayer's efforts and his friendship with Professor James Douglas and his son James Stuart "Rawhide Jimmy" Douglas who developed the United Verde Extension Mine in Jerome, Arizona starting in 1912. At the time the pipeline was constructed, the Douglas' were living in Mayer. The *Prospect* reported

Eventually, the Mayer Apartments were converted to one apartment per building by cutting interior doors through from one unit to the next. In 1989, the Mayer Apartments were listed in the National Register of Historic Places. Photograph by Nancy Burgess, 1989.

in 1904 that "there is completed about eight miles of pipeline, from ever-flowing springs of pure clear mountain water, with a capacity of 400,000 gallons per day, which will be sufficient for a town of at least 6,000 inhabitants."

Building Mayer wasn't Joe Mayer's only interest. He had interests in numerous mines in addition to the Onyx Mine, which he co-owned at one time with William O. "Buckey" O'Neill, Prescott's famous Rough Rider, and at one time was managing the Henrietta Mine near Mayer. He had a number of mining claims on his own property right in Mayer, including the "Winneford," "Martie No. 2" and the "Jickey No. 2." He did not consider himself to be a miner and was not interested in the intensive labor necessary to actually mine, but he was a very interested and active mine investor. In addition to all of the grubstakes he provided, he purchased shares in numerous mines in Yavapai County, as was the custom for almost every serious businessman in the county. Investing in mining claims, equipment and development companies was standard procedure for Joe Mayer. His wife, Sadie, also owned mining claims.

Another of Joe Mayer's enterprises was an experiment, along with E. S. Rogers, in producing cactus toothpicks. On August 27, 1902, *the Arizona Weekly Journal-Miner* ran an article about the enterprise, stating that Mr. Rogers interested Joe Mayer in the idea of producing toothpicks from cactus spines and the two began a series of experiments to find a method for treatment of the cactus spines for toothpicks. The article further described the process for preparing the cactus spines for their transformation into toothpicks:

The Mayer onyx mine, in which Joe Mayer had an early financial interest, about 1912.

First he goes out and gathers a load, bringing them to his little workshop, with a pair of big shears, such as tinners use, he shears off the thorns. Then they are treated to various chemical baths, thousands of them at a time, from which they emerge brilliantly colored through the removal of the outer surface, yet rendered even more pliable than before, instead of more brittle, as might occur if not treated. The points are then dulled by machinery and they are ready for boxing and sale as "Indian Souvenir Toothpicks."

A September 12, 1902, follow-up article in the *Arizona Weekly Journal-Miner* gave an update on the cactus spine toothpick enterprise:

Under the name of the "Indian Souvenir Toothpick Company" a plant is in operation at Mayer that shows the natural resources of this section in so far as furnishing toothpicks go. Messrs. Mayer and Rogers are the men who are behind the enterprise, and are now diligently employed in not only preparing the article they semi-manufacture, but in marketing the same. They

This cabinet card photograph of Sarah Belle Mayer was probably taken after Joe Mayer's death about 1910. It was identified and donated by Winifred Mayer Thorpe in 1978 (courtesy Arizona Historical Society/Tucson no. 61053).

are up to date unable to supply the demand owing to the great number of orders received. The little instruments they are handling are the production of the cactus plant, serviceable in their duty more so than the common wood article, and when polished are variegated in color. Mr. Rogers, who "caught on" to the idea of preparing the thorn from the plant, has labored earnestly to prepare a chemical solution to strip the branch of its natural growth and has succeeded admirably. The toothpick he produces is about two and one half inches long, has an amber tip of natural color, while among the stem is a variety of colors, showing the natural condition of the plant. There is nothing whatever artificial in the toothpick in either color or composition. They are also durable and will give better service than the manufactured goods. A box may be seen at this office, and a sample will be given any who may call.

Although the *Arizona Weekly Journal-Miner* reported that Mr. Rogers had patented the process he used to prepare the toothpicks, no records could be found of this process and no further information on the fate of the business was located. Although the newspaper claimed that the cactus spine toothpicks could be sold "as cheap as a good quality of wooden toothpicks," the business seems to have faded away. Perhaps the process was just too labor intensive and expensive to make it profitable.

Joe Mayer, George Byron Scammel and Edwin D. Treadwell partnered to form the Mayer Realty and Investment Company, which was incorporated in the spring of 1904. The Board of Directors consisted of Mayer, Treadwell and Scammel. Joe Mayer was President and Treasurer, Treadwell was the Vice-President and Scammel was the Secretary of the corporation. The purpose of their company was stated in their Articles of Incorporation as engaging "in the purchase of real estate, marketing the same, the construction of such railroads, tramways, pipelines and buildings as may be deemed necessary, the development of water and controlling the same, the bonding and purchasing of mines, mining and other stocks, to do a general real estate, mining and stock investment business, to

In January 1905, an eye-catching advertisement for the Mayer Townsite ran in the **Arizona Gazette.** *The advertisement pretty much covered everything a prospective buyer needed to know about Mayer—"The Coming City of Northern Arizona, the Center of the Mining, Pastoral and Agricultural Industries of Yavapai, Arizona's Richest County." Lots, 50 × 150 feet were "$100 to $500, terms cash with discount or one-third cash, balance six and twelve months, 8% interest. Lots sold and houses built on installment plan. Tent City for those who prefer it. Everything arranged for health and comfort."*

sell, exchange, lease and mortgage all properties that the corporation may acquire." On November 22, 1904, they recorded the plat for the Townsite of Mayer, Yavapai County, Arizona, consisting of approximately 89 acres. It was surveyed by B. L. Smith. The land was subdivided and platted with streets, alleys and lots. The property was located south of Joe Mayer's properties on Central Avenue in downtown Mayer. Street names included Mountain View, Terrace Drive, Yavapai Avenue, Agua Fria Avenue, Fair Mist Avenue, Big Bug Avenue, Arizona Avenue and Mayer Place. Cross-streets were labeled First Street through Ninth Street. Underlying the subdivision lots were patented mining claims, including the Treadwell, Old Soldier, Home, Fair Mist, Winneford, Martie No. 2 and Jickey No. 2. As is customary when platting a subdivision, the "avenues, streets, parks, plazas, public grounds and alleys" were dedicated to the public for their general use and benefit. Some of the street names still exist and others have been changed or no longer exist. Part of this subdivision is now well developed, but the southwestern corner of the subdivision has never been developed.

The March 1918 issue of *Yavapai* magazine ran a lengthy article encompassing several pages on "Mayer—the City With a Future." In reporting on Mayer's earlier days when Joe Mayer was "the man" in town, the reporter wrote: "The growth of Mayer may conveniently be divided into two periods—the past and the present. A few years ago Mayer was one of those real Western towns of the saloon and the cowboy sort. The chief pursuits of the men of these days were cattle raising, mining, poker and occasional gun plays. For over thirty years Mayer neither progressed nor prospered as progress is counted now-a-days. It enjoyed a happy, dreamy standstill, with brief, long-separated intervals of activity. For thirty years, Joe Mayer held all the real estate and, though he encouraged every good movement that meant the good of the community, he somewhat retarded the growth of the country by holding his many acres when he was not able to improve or develop them. Such was the past."

Joe Mayer would not have agreed.

7

Business Is Booming

Who would imagine that the merchants of Yavapai County were the backbone of the economy in the early years of the development of the Territory of Arizona? But they were, and the rise and sometimes fall of the prosperity of any small town in the Territory could be linked, one way or another, to the town merchants. In the early history of the Territory, most merchants relied on the capitalist system, particularly in rural areas. Where cash money was in short supply and banks were few or nonexistent, barter became the predominant method of buying and selling goods and services. As Scott Fritz wrote in an article published in the *Territorial Times* Vol. 1, No. 2, entitled "Yavapai County Merchants: The Center of Arizona's Early Economy, 1863–1881":

This photograph of the Mayer Business Block taken about 1905–1906 shows a group of men in front of the Mayer Saloon. The man in the background with his hand on his hat is Joe Mayer. The man in the white apron is probably from the restaurant next door. Notice the rounded display window, which has a matching mate to the right. Unfortunately, this part of the curved window no longer exists, but the other half is still intact.

This type of economy prevailed in rural areas of nineteenth century America with small populations with little cash and where stores served as banks. It was an economy based on barter where producers of raw products obtained tools and seed on credit and paid store debts with commodities they produced. Merchants sold the raw products to local and distant markets, extinguished their customers' debts and provided them more credit or currency. In Yavapai County, farmers, stock growers and miners paid their bills with gold, lumber, wheat, corn, meat and forage. Merchants sold agricultural products to nearby forts and bullion dealers outside the territory, thereby liquidating their debts as well as those of their customers and earning a profit on the difference.... Yavapai County's economic history was microcosm of Arizona's overall economy and showcased the importance that merchants played in Arizona history from 1864 through the 1870s.

And, as Gail Gardner, son of an early Prescott merchant, wrote in an article entitled "The Pioneer Merchant," published in *Echoes of the Past, Tales of Old Yavapai Arizona*, Volume 1, "Your pioneer merchant kept a simple ledger and cash book himself, and did not need the services of a certified public accountant or a tax attorney. But the pioneer merchant was secure in the integrity of the old time cowman; when the steers were sold, the cowman paid his bill and that was that. A man's word was good and no scrap of paper could make it any better."

James I. Gardner came to Arizona Territory in 1879. He built his new mercantile store in Prescott in 1890. Typical of mercantile stores at the time, Gardner's carried just about everything. Gardner also took mail orders, which he put on the train every afternoon. The S F, P & P would drop off the orders along the route of the railroad for pickup by the customer (Sharlot Hall Museum).

7. Business Is Booming

A typical ledger sheet for the Mayer Livery, Feed and Sales Stables, run by William L. "Bill" George. The customers are the Stewart Brothers of the T Anchor Ranch. Bill George was a stagecoach driver who later worked for the P & E at the Mayer Depot (Larry Howard).

This outstanding photograph of Mayer is taken facing southwest about 1900–1901. The front of the Mayer Hotel is on the left. A Prescott & Eastern locomotive, tender, combination car and four passenger cars had just crossed the small trestle on the way to the Mayer Depot. The back of the Mayer Station (with the bay window) includes the back yard and a small barn, which is still on the property. Four ladies are walking down the street toward the Mayer station. A pile of lumber where the Mayer Business Block would be constructed indicates that construction may be started soon (courtesy Arizona Historical Society/Tucson, no. 61055).

Telephones were an important amenity in a small town such as Mayer. The first "telephone directory" for the Mayer/Humboldt area was published in 1902. This 1917 Yavapai magazine advertisement for the Mountain States Telephone and Telegraph Company promoted the value of the reasonable rates for telephone service versus the value of other commodities which had gone up in cost.

Starting in 1902, with the "Independent Telephone Directory," the community directories gave a good idea of how Mayer was progressing. Of course, initially, very few businesses had telephones and almost no households had them. From the 1902 telephone directory there is a list of mining camps and towns connected together by telephone. The caller is instructed to "call for these stations by name." Apparently, those listed actually had telephones on the premises and local residents could be "fetched" to come to the telephone if there was an incoming call or go to the specific business to make a call. All of these calls were long-distance. Included in Mayer in 1902 were Grant Bros. and Joe Mayer's. By 1907, the popularity of the telephone had made a significant difference in the listings for Mayer and vicinity and included some residential listings. Businesses with telephones included the Mayer Hotel, the P&E Depot, Joe Mayer's Store, the Mayer Hospital, the Rigby Reduction Works, the Treadwell Mining Company office and the Treadwell Smelter, Fred Venator's Market and Fred Venator's Slaughter House, Yavapai Supply Company, the W. E. Frazer Lumber company, Trice & Morgan and the Wagner Hotel.

Additional information about the progress of Mayer can be gleaned from the 1910 United States Census, which listed three retail merchandisers in Mayer: Wilbur Mayer, Henry Heffleman and Charles E. Trice. The census also listed a commercial lumberman, James B. Elder, and a hardware salesman. Other businesses represented in the census included a seamstress, a Chinese laundryman, a Chinese restaurant owner, three blacksmiths, a shoemaker, two music teachers, several farmers and stockmen, seven "house" carpenters and a paperhanger, a meat shop proprietor and three saloon keepers. Obviously, the little town of Mayer was growing up and had become an important center for commerce in the area.

Joe Mayer was the town of Mayer's first, and probably its best known, merchant. Most of the retail businesses in Mayer from the time the Mayer family arrived in 1882 until Joe Mayer's death in 1909 were under the name of "Mayer." The Mayer family apparently stayed in business in Mayer until at least 1950, as shown by an advertisement in the 1950 Mayer High School yearbook, *La Loma del Cobre*, for the "Nelson Tavern and Amusement Hall." Wilbur (Burr) Joseph Mayer died on December 27, 1955. His son Wilbur Nelson Mayer, who ran the Mayer pool hall in the Business Block and was Burr's partner in the business of Mayer & Mayer, predeceased him, dying on April 22, 1952.

Joe Mayer established the Mayer Mercantile almost as soon as he arrived at Big Bug Station. Aware that there were no other merchants in the vicinity, he saw an opportunity to serve the mining and ranching interests of the area. His mercantile was originally housed at the Mayer Station and later, (after 1897), in the Mayer Hotel. While the business was located in the Mayer Hotel, it was managed by C. E. Wicks. Later, after the mercantile store moved across the street to the Mayer Business Block in 1902, the name of the enterprise for a time became Wicks & Mayer. Wicks & Mayer, which was generally known in the area as "Joe Mayer's Store," carried general merchandise and mining supplies. As Winifred (Winnie) Mayer Thorpe wrote in her article "Joe Mayer and His Town," Joe's main fault was that he was too kind hearted and generous, that he could "never say no after hearing some hard-luck story and would extend credit where it was not deserved. Many took advantage of his generosity,

GENERAL MERCHANDISE

Dry Goods

Groceries

Stockmen's and

Miners' Supplies

Our Stock Is Complete and you will find that Our Prices Are Reasonable

Years of experience has taught us what our patrons want and we try to keep it on hand at all times. If you deal here we know that you will be satisfied. Our ever-increasing business is the best of evidence that we treat our customers right. We solicit your patronage, guaranteeing satisfaction.

Come In and Get Our Prices
Compare Our Goods With Others

W. J. MAYER

Mayer, - - - Arizona

A newspaper advertisement for W. J. Mayer states: "Years of experience has taught us what our patrons want and we strive to keep it on hand at all times. If you deal here we know that you will be satisfied. Our ever-increasing business is the best evidence that we treat our customers right."

An interior view of W. J. Mayer's store. The animal pelts from an earlier time are gone as are the Indian rugs, however, the beautiful Yavapai basket is still on display on the counter to the right. The merchandising has changed, and the store shows off a substantial selection of bolts of fabric and dry goods. Burr Mayer is standing on the left.

borrowing large sums of money which were never repaid. Others moved away and left no forwarding address. Some put off paying until the debt was outlawed. But always Father's faith in mankind was very great." Consequently, the store was usually "in the red." Winnie also wrote that once the railroad arrived, "it was mostly cattlemen who kept business good in the store" and that "often there would be three or four thousand dollars on the counter in payment for their provisions." Some of Joe's worry about the financial situation and the running of the store was relieved when Burr Mayer, his wife Annie and their infant son, Joseph, moved back to Mayer in about 1904 and Burr took charge of the store. At the time of Joe Mayer's death in 1909, the Mayer Mercantile was in serious debt due to too much credit given and too few collections realized. In order to put the store back on an even keel, Sadie Mayer made the decision to destroy the old ledgers and start fresh, which Burr accomplished, as Winnie wrote, "with a heavy heart." Burr Mayer then set about the task of clearing the business of all debts and Sadie turned the business over to Burr to run as his own.

The White House Hotel, originally called the "Hotel Wagner," was purportedly built in 1903 by Frank Wagner (or Wagoner) to serve the large number of ranchers, miners and mining men who frequented Mayer at the time. By 1909, it was owned by Mrs. H. B. (Louise) White, who changed the name to the "White House" Hotel. There were nine rooms in the hotel and a bunkhouse and cabins out back for the overflow. When the weather was nice, the guests slept out on the upstairs porch. In

The "Bunk House" behind the White House Hotel is still standing, although it is now used for storage. The hotel is no longer open, but the owners enjoy it as a retreat from the Phoenix area. Photograph by Nancy Burgess, 2011.

1912, the Louis Schrade family of Switzerland bought the hotel. Louis Schrade was a miner. His wife ran the hotel. They kept the name and almost everything else just the same as it had been under Mrs. White. The March 1918 issue of *Yavapai* magazine said of the White House: "The traveler, who is particular, can put up at the White House and know that he has secured about the best in Mayer, which is as good as one can find elsewhere in the State." In her letters home to friends and family, Mrs. Scrade wrote that the only help they could get were Chinese, but that they were not reliable because as soon as they could, they went back to China. Mrs. Schrade asked for girls to come from Switzerland to make beds and clean the rooms. Martha Hegglin was crazy about horses and the "wild" American West. She read everything she could find about cowboys and Indians, Annie Oakley and Buffalo Bill. She answered Mrs. Schrade's plea for "reliable" girls to work at the hotel and when she was almost 19, in

WHITE HOUSE
NICE, COMFORTABLE, HOME-LIKE HOTEL
P. & L. Schrade, Proprietors
MAYER, ARIZONA

A 1922 advertisement from the **Big Bug Breeze** *for the White House Hotel mentions the owners at the time, Mr. and Mrs. Schrade.*

7. Business Is Booming

Rosena Promberger Minucci took this photograph in 1941 of the White House Hotel. Neither the setting nor the building has changed much since this photograph was taken, just the models of transportation.

1913, she and her cousin, Marie Brandley, came to Mayer. Martha's sister, Josie, who had encouraged Martha to come, was already working at the White House Hotel. In an interview published in the July 26, 1974, *Courier*, Martha Hegglin Hickey said that "coming from a small town in Switzerland, I thought I had reached the end of the world when I came to Mayer. America is so vast and Switzerland is about the size of Yavapai County." Mayer was booming at the time with mines such as the Blue Bell, DeSoto and Swastika going full blast. Martha said in her interview that they "were busy all the time in those days. The place was always full of miners and cattlemen and the place was jumping all the time." The miners "took their board there," keeping long-time cook at the White House, Agnes Zurcher, "flying around in the kitchen, cooking for at least two settings of miners three times a day and keeping the oval dining room table, known as the 'society table' piled with aromatic food." The March 1918 issue of *Yavapai* magazine further said of the White House: "The White House is a two story brick building, of 16 rooms, surrounded on all sides by a spacious, vine-covered veranda, which makes it especially pleasant in summer. An excellent table is kept in connection with the hotel." The boom continued through World War I, but once the war was over, the demand for metals dropped significantly, and the busy days of the White House Hotel were numbered. Agnes Zurcher would later own the hotel. Eventually, in 1951, Martha Hickey's son William bought the hotel for her to run. Preservation was Martha's goal and every room in the hotel was a museum. On the back porch were the three phases of refrigeration used by the hotel—the 1903 Arizona Cooler, a later Polar King Ice Box, and a huge walk-in Frigidaire refrigerator. The Arizona Cooler was a screened cupboard upon which cool water dripped continuously.

MRS. MARY L. WELLS

DEALER IN GROCERIES AND DRY GOODS

Mayer, Arizona

Another 1922 advertisement from the **Big Bug Breeze** *is for grocer Mary Wells. Mrs. Wells was one of the few female business owners in Mayer at the time and she was in business for a number of years.*

THE JOE MAYER CLUB

WM GILMER

Best Brands of Wines, Liquors and All Leading Brands of Cigars

Courteous Treatment To One and All

Best Soft Drinks   First-Class Pool Tables

Stockmen's Headquarters.   Your patronage appreciated.   Mayer, Ariz

---

A 1916 advertisement for the "Joe Mayer Club" for wines and liquors, is a mystery, not only because it seems to be the only reference to the "Joe Mayer Club," but because Arizona went "dry" on January 1, 1915. It is possible that the proprietor, W. M. Gilmer, thought that by advertising as a private club, he could get around the requirements of Prohibition. No profession is listed from Mr. Gilmer in the 1916 Directory, and, by 1917 he is not listed at all.

Martha said that "just a little bit [of water} dripped onto a wet wool sock" kept things pretty cool in the Arizona Cooler, although food did spoil sometimes. In the fall of 1976, Martha was forced to close the hotel. A landmark in Mayer, It is now owned by Mayer natives Clyde McDonald and Sonya Hickey McDonald.

As Mayer was growing up after the turn of the century, new entrepreneurs started businesses in town. Some were newcomers, others were long-time Mayerites or lived at the nearby mines or on ranches in the area. Just as Joe Mayer had, they saw an opportunity and made the best of it. In spite of a national depression in 1907, by 1913 the Yavapai County Directory devoted almost two full columns to Mayer. Businesses included the Arctic Ice and Meat Company; Joseph Cook's saloon and pool hall; Ish

Davis' saloon; the Mayer Hotel, which was no longer in the ownership of the Mayer family; the *Mayer Miner* newspaper, edited by J. E. Hill; and Mayer & Mayer, under the ownership of Burr Mayer. Also, the Mayer Water Company; the Rigby Mining and Reduction Company; the White House Hotel; Wells Fargo & Company Express; Mrs. Mary Wells, grocer; Western Union; and Yavapai Supply Company, owned by Henry C. Heffleman, were listed. Of Heffleman's hardware store, the March 1918 issue of *Yavapai* magazine said: "The store, under the management of H. C. Heffleman, locally known as 'Hank,' has been of invaluable service to the consumer needing odds and ends in the hardware line. A complete stock of hardware supplies is kept." New businesses reflected well on Mayer, but it did not grow to any extent from 1907 until shortly before the start of World War I. Mining and ranching kept the town going, with a steady population of around 600, but few new projects were started during this period.

A growing community brings with it amenities of civilization: post offices, churches, libraries, schools, music teachers, doctors and hospitals, banks, courts, sports facilities, utilities such as electricity and telephones and law enforcement. All of these eventually came to Mayer, most of them in the early 20th century. The years just before World War I and during World War I resulted in a lot of growth and prosperity for Mayer and its residents and business owners. The metals market was doing very well and mining was at its peak in the area. An article in the March 1918 issue of *Yavapai* magazine, entitled "Mayer—The City With a Future," said of Mayer:

> Superficially, the town of Mayer looks very much like a summer resort to the casual visitor. It is beautifully situated beneath giant cottonwoods on the level valley floor of Big Bug Creek and completely surrounded by lofty pine covered mountains. But, of course, Mayer isn't a summer resort. It probably will be when Arizona gets educated to it. It is still now, in good part, as it was in the early days, a provisioning post for cattlemen and sheepmen. It is however, rapidly becoming a mining center....
>
> In fact, mining is now the big thing. The mines adjacent to Mayer have ushered the community out of its first period and set it down in the middle of the second period of its growth....
>
> The striking thing about Mayer's growth is that it lacks the mushroom qualities that most mining towns have. All of the new business buildings are substantially built, and give the town and air of permanence. The new residences are, for the most

In 1913, one of Mayer's businesses was Joseph Cook's Saloon and Pool Hall. It wouldn't be too long until Arizona went "dry" and gloom would settle over the drinking set. Prohibition theoretically shut down the saloons, which would require Mayer's several bars had to find other custom. This trade token advertises Joe Cook's, and was good for 12½ cents (one "bit") in trade.

part, large, well-built and painted....

All of this activity indicates and era of prosperity for Mayer and very probably the bringing in of several new mines in the district.... This means that Mayer will be the headquarters in every sense of the word of one of the big mining districts in Arizona....

During the last year, most of the merchants have responded to the wave of prosperity that is beginning to be felt in Mayor [sic], either by enlarging their stores or by adding new departments to them.

The 1917 directory listed a population for Mayer of 700. In 1919, the population was listed in the directory as 1,000. Several new professions and businesses were established in Mayer during the 1917–1919 time period of World War I. One "new" profession listed was "stage driver" for the "auto stage." At this time, a "stage driver" or "auto stage" operator was someone driving a motorized vehicle for hire rather than a stagecoach. However, there was still a livery stable in Mayer and there were four blacksmiths. In fact, Mayer continued to have two to four blacksmiths working in the community through the 1930s. The last listing for a blacksmith in Mayer was in the 1937 directory. Also around this time, the first vocations which reflected the trend away from transportation by horse to the automobile are listed in the directories. In an interview in June of 1981 for the *Prescott Courier*, Winifred Mayer Thorpe talked about "the day horses became obsolete." She said that "a man came through town in a funny-looking thing and he wanted Dad to ride with him. He said after awhile there wouldn't be any more horses. I wanted to go so bad, but I was afraid to ask." But she remembered asking her father how the ride was. Joe Mayer said, "Winnie, it was just like riding in a wheelbarrow." "Horseless Carriage" related businesses really began to proliferate around 1919 with the establishment of the Mayer Garage by Harry McMichael (the same man who was the "picture operator" at the Mayer Owl). Also listed in the 1919 directory were a truckman, auto livery, a garageman, two mechanics, a master mechanic and a chauffeur. As automobiles and trucks became more common, Mayer boasted quite a number of businesses which depended on the "machine" for their business. Burr Mayer advertised Goodyear tires on his business sign for W.

Ride The Jordan Stages

FOR
Service, Safety and Comfort
TO
Dewey, Humboldt, Mayer, Henrietta, Stoddart and Blue Bell Mines.

SPECIAL SERVICE
TO
Cherry Creek, Camp Verde, Cottonwood and Jerome

Leave Prescott for Mayer
9 a. m. and 3 p. m.
Leave Mayer for Prescott
8 a. m. and 2 p. m.
Leave Prescott for Jerome
9 a. m. and 2 p. m.
Leave Jerome for Prescott
9 a. m. and 2 p. m.

PRESCOTT OFFICES:
Scholey's.   Phone 108
Bruchman's. Phone 411

Echert & Wilson
Props.

By the date of this 1918 advertisement, "stages" were running regularly from Prescott to Mayer and back. Note that it took six hours to get to Mayer from Prescott by vehicle and the same amount of time for the return trip. The same trip on the P & E took about the same amount of time.

7. Business Is Booming

This exceptional postcard image was taken from the southeast looking northwest. The Black Canyon Road runs from center right to center left. The P & E Depot is on the far right with S F, P&P boxcars sting on the siding. The Mayer Owl (light roof) is clearly visible across the street from the P & E water tank as is the side of the Mayer Hotel. The most interesting detail, however, is the very early truck in the foreground which is being driven by a man wearing a hat. There is another man standing in the bed of the truck. He is holding on to whatever is being transported. Could this be similar to the "funny-looking thing" in which Joe Mayer went for a ride in 1909? The postcard was mailed from Mayer in 1917. The message reads, "This is Mayer some time ago. You should see it now."

This invoice dated January 1, 1921 from the Mayer Garage lists interesting prices for repairs made to rancher Fred Dugas' vehicle, which also included an inner tube for $6.00. Labor was $1.25 an hour (Larry Howard).

Alabam Freight Lines wood transfer trucks are shown lined up at their yard at 418 Moeller Street in Prescott in May 1934. Alabam Freight made regular runs to and from Mayer with deliveries in the 1930s (Sharlot Hall Museum).

This bright yellow and black metal sign advertised Goodyear Tires at Burr Mayer's store.

> **MAYER AUTO COMPANY**
> Auto Supplies Supplied.   Opposite Post Office
> L. B. PRICE, Proprietor
> MAYER, ARIZONA

The Mayer Auto Company, owned by Louis B. Price throughout the 1920s, advertised in the **Big Bug Breeze** *in 1922. This business was located next door to the Mayer Hotel and later became the "Highway Garage." A small storage building now occupies the spot.*

Frank Garrett apparently arrived in Mayer about 1931. He ran the Highway (or High Way) Garage for many years. Mr. Garrett posed out in front of the garage in 1941 next to a manual "visible" gas pump flanked by two more modern "Flight Gasoline" meter pumps. Bill Promberger worked at the garage starting when he was about 12 in the mid-1930s until he graduated High School in 1941. His brother-in-law, Archie Minucci, was a mechanic at the Highway Garage. The building was destroyed in the snow of 1967.

The Mayer Owl began business about 1918. Alfred E. Rice owned the Owl, a combination drug and sundries store with a soda fountain, restaurant and movie theater. Frank Promberger came to Mayer in 1923 and managed the Owl for Mr. Rice, whose occupation was listed in the directories throughout the 1920s as "miner." This interior photograph about 1925 shows the soda fountain along with all sorts of merchandise, including pipes, pocket watches, greeting cards and Cracker Jacks, piled up behind the bar. Mike, the cook, is on the left, then Mr. Rice, then Elsie the waitress and Frank Promberger on the right. The building burned in 1931 (Bill Promberger).

J. Mayer in about 1917. Through the 1920s, more "mechanics" and "garagemen" and fewer blacksmiths were listed. In the 1931 directory, Frank J. Garrett appeared for the first time, listed as an "auto dealer." Garrett would later own the Highway (or High Way) Garage on Central Avenue between the Mayer Hotel and the Mayer Apartments. Bill Promberger worked there starting when he was about 12 years old (1935) until he graduated from high school in 1941. His brother-in-law, Archie Minucci, was a mechanic and also worked at the Highway Garage. The building collapsed in the great snowstorm of 1967–68.

The "one bit" trade token advertised the Mayer Owl on the reverse side of the token, rather than the front.

Trade tokens were sold to merchants by traveling salesmen who showed samples to the customer, who could then choose the shape, size, finish and design to order for his or her business. They were the most popular with saloons and restaurants. This eight-sided pierced token has a value of "one bit."

The Mayer Owl was one of the businesses listed in the 1919 directory. Regarding the Mayer Owl, the March 1918 issue of *Yavapai* magazine states: "One of the new businesses that is rapidly growing popular in Mayer is the Mayer Owl. Here are kept a line of drugs, periodicals, fountain specials, candies and tobacco. Recently, special arrangements have been made whereby a splendid restaurant service has been added. On Tuesdays, Thursdays and Sunday nights motion pictures are shown in the auditorium, in the rear of the store. Occasionally, dances are given here also. The firm owns and operates its own electric plant, thus insuring against light and power shortage." This was the first mention of motion pictures in Mayer and the "picture operator" was J. H. M. McMichael. In 1923, Frank Promberger and his wife, Caroline (Carrie), and their three children came to Mayer from Arkansas for Mr. Promberger's health. The Prombergers managed the Mayer Owl along with Alfred E. Rice, whose occupation until 1929 was listed as "miner." After Mr. Promberger passed away in 1930, Carrie continued at the Mayer Owl, and the family lived in the building. Unfortunately, the Mayer Owl burned to the ground on June 28, 1931. Consumed in the fire were the Mayer Owl, the Lev Nellis' Butcher Shop, the Arctic Ice Plant, a four-room house owned by Mamie Mayer and the Sam Lee Chinese Laundry. The family, consisting of Mrs. Promberger, her older daughter, Rosena Promberger Minucci, Rosena's husband, Archie Minucci, and Mrs. Promberger's younger children, lost everything in the fire and were fortunate to have gotten out of the building in time so that no one was injured. The volunteer fire department worked hard to keep the fire from spreading to

The Promberger family is photographed in front of the Mayer Owl in 1923, shortly after their arrival in Mayer. Included are Rosena, Frank, Helen, Carrie and Bill Promberger. The woman on the right is unidentified. The building was located just south of the Mayer Business Block. Oak Street separates the location of the two businesses.

THE MAYER STATE BANK

CAPITAL STOCK $ 25,000.00

The Mayer State Bank letterhead of 1924 states that the capital stock was $25,000.

the Mayer Business Block across the street as pieces of tar paper from the roofs of the burning buildings spread up to 1/4 mile away. After moving back to Arkansas for a time, Carrie returned to Mayer in 1934 where she and her son Bill lived with Sadie and Mamie Mayer in the old stage stop. Her daughter Helen lived with Wilbur, Annie and Nelson Mayer. Carrie Promberger cared for Sadie Mayer, who was bedridden or confined to a wheelchair, until her death on November 11, 1934, and for Mamie Mayer until Mamie's death on January 10, 1964.

An article in the December 1917 issue of *Yavapai* magazine reported on the new Mayer State Bank, stating: "The advantage of having a bank of their own appealed to the people living in and about Mayer early in 1916. In a short time after the opening of the bank, its books showed that almost all the cattle ranchers and mining operators in the vicinity of Mayer were depositors." *Yavapai* magazine further reported in the same article that the institution was established by R. C. Walters in April 1916 and that, at the time of organization, the officers were president, Earnest LaDue; vice-president Charles Batre; cashier, R. C. Walters; and assistant cashier, E. A. Fisher. According to the June 1917 issue of *Yavapai* magazine, the Mayer State Bank opened April 7, 1917. The March 1918 issue of *Yavapai* magazine said of the Mayer State Bank: "The Mayer State bank is housed in one of the most substantial buildings in Mayer. It is due to the efforts of Mr. R. C. Walters that the bank was established, and it is due to his management since its establishment that the bank has steadily grown.... When the bank was opened, a little more than a year ago, the books showed a very modest number of depositors. Today, however, the number of depositors have increased substantially. The Bank of Mayer is closely following the growth of the district." The December 1919 issue of the *Big Bug Copper News* listed the assets of the Mayer State Bank as $259,501.85 and liabilities of the same amount. Unfortunately, the Bank of Mayer was short-lived. The Prescott State Bank closed on November 25, 1925. It can be assumed that the fate of the Mayer State Bank was similar, as by 1926 Mr. Walters was no longer in Mayer and the bank was no longer listed in the directory. It would be decades until Mayer again had a bank. Later, the Mayer State Bank building would house a variety of businesses including a drug store with an ice cream shop. It eventually became the home of Frank Polk, cowboy, rodeo performer, Western artist, author and all around "character." Since 2003, a florist's shop has occupied the Mayer State Bank building. Many of the original features of the bank are still in place, including the large Diebold safe, which is on display in the florist's showroom.

Sadie Mayer was well known in the early days of Mayer as the town "doctor." Although she was not a doctor in the professional sense, she was very good at taking care of the ill and injured, and the Mayer Station was always open to those who needed her care. Ranching and mining are dangerous professions, and injuries were common.

The Mayer State Bank was very important to a community that had never had a local bank and customers flocked to it when it opened in 1917. This snapshot was taken shortly after the bank opened under the management of R. C. Walters.

According to Winnie's article "Joe Mayer and His Town," Sadie once drew a sliver of metal from the eye of a miner with a magnet. Further, much of Yavapai County was well promoted as a haven for "health seekers," particularly those who had diseases of the lungs. Around the turn of the century, there were at least 13 sanitariums in Prescott. Joe Mayer himself promoted Mayer as having the "finest climate in the world for debilitated people who can and want to own their own homes." Although Mayer did have an on-again, off-again hospital, the large hospital closest to Mayer for many years was Mercy Hospital in Prescott, 25 miles away. In the 1920s, a hospital in Humboldt provided services a little closer to Mayer. Dr. Robert N. Looney, Joe and Sadie Mayer's son-in-law, came to Yavapai County in 1896 and ran a small hospital at McCabe and was also the physician for the mines at Crown King until 1905. After he and Martie Mayer were married, they moved to Prescott. He was the first state health officer for Arizona from 1912 to 1917. Although he could and would take the

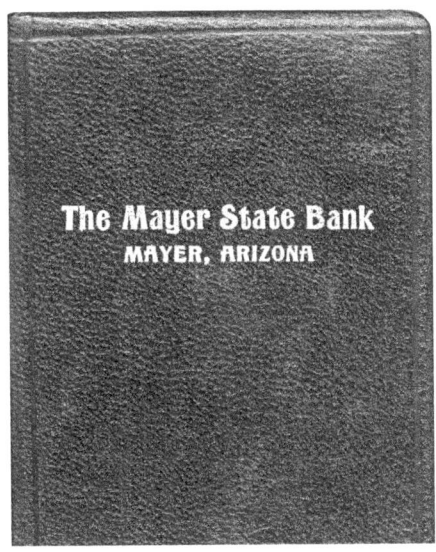

This is a photograph of a Mayer State Savings Bank. Manufactured by Bankers Utilities Company, it is covered with leather and is made to look like a book. It is 3½" by 5" with a slot at the top for cash. The box for the bank says, "We keep the key. Bring this savings bank in at least every thirty days to have it unlocked and the contents placed to your credit so your saving will begin to draw interest."

After serving as a home for Frank Polk for many years, the Mayer State Bank building is currently occupied by a florist's shop, "Lynn's Foliage Bank." The huge Mayer State Bank Diebold safe is a centerpiece of the shop. Photograph by Nancy Burgess, 2009.

Many people came to Arizona, and particularly to Yavapai County, seeking relief from respiratory illnesses, including Frank Promberger. This photograph, captioned "health seeker" circa 1930, shows an inventive traveler along Route 66 in Arizona with a motley crew of at least nine dogs. Hopefully, he and the dogs found what they were seeking.

Mercy Hospital in Prescott was founded by the Sisters of Mercy in 1898. It had doubled in size by the time this photograph was taken. It was the only large, well staffed hospital in the area and served all of the regional communities including Mayer. Joe Mayer's son-in-law, Dr. Robert N. Looney, would have treated patients at this hospital. The building burned in 1940.

train from Prescott to Mayer if needed, he apparently never established a practice in Mayer.

The Mayer Hospital seems to have a sketchy history. The city directories do not list a hospital, nurse or doctor until 1931. An obituary from the *Prescott Journal-Miner* dated February 3, 1909, reports the death of Dr. E. A. Hall of Mayer who "came to Mayer four years ago and established a hospital there." At the time of the Spanish influenza epidemic, which broke out in Prescott in October of 1918, there appears to have been a hospital of some type in Mayer, as a photograph shows Charles Kinsman and Frank Wilson recuperating from the flu in Mayer. Apparently a victim of the post–World War I economic downturn, the hospital seems

The "old" Mayer Hospital, on Main Street near Pine, was supposedly built in 1902. With broad porches to accommodate wheel chair patients, it is typical of buildings constructed of wood in Mayer at the time. It apparently closed around 1919 and was later converted to a private residence.

This photograph of a stately brick building in Mayer is labeled "Sanitarium." Newspaper articles do mention that patients were being treated in Mayer for tuberculosis, but no other details were provided.

to have closed around the end of World War I. Photographs from 1920 show the "old" hospital as having been converted to a private residence. The December 1919 issue of the *Big Bug Copper News* reported that "the Mayer Hospital Association has been organized by the citizens of Mayer and vicinity. It is the intention of the association to raise about $15,000 for the erection of a modern building.... The management will be under the care of Dr. James B. Van Horn, M. D. Dr. Van Horn has been recently discharged from Whipple Barracks, Prescott, Ariz., coming there in late May 1918 and serving in the medical corps until September, 1919.... The citizens of Mayer should be congratulated on having in their midst an able physician and worthy citizen such as Dr. James B. Van Horn." Although Dr. Van Horn is not listed in any of the directories for Mayer except 1923 and 1925, he is listed in the 1920 United States Census as a physician at Mayer Hospital, age 37, with a wife and son. However, it wasn't until two years after the announcement that Dr. Van Horn would be arriving in Mayer that Articles of Incorporation were published in the *Prescott Journal-Miner* on January 30, 1922, for the "Mayer Hospital Association, Incorporated." The purpose of the corporation was to "build, erect, lease or to otherwise acquire, and to maintain, a hospital and sanitarium." Further, on December 12, 1922, John A. Martin, a Mayer businessman, sold lots numbered 10, 11, 12, 13, 14 and 15 in Block 2 of the Mayer Townsite to the Mayer Hospital Association. These lots were on Fair Mist between First and Second Streets. A photograph, probably taken in the 1920s, of a brick building in Mayer is labeled "Sanitorium." The 1931 directory lists a Dr. Warren Baldwin and a nurse, Mrs. Julia Adney, in Mayer. Mrs. Adney is listed in the directory through 1937. Dr. Baldwin's name does not appear again after the 1931 entry.

The post office had been established in Mayer in 1884 with Sarah Belle Mayer

The Mayer Post Office, built in 1917 next door to the Mayer Business Block was apparently owned by Sadie and Mamie Mayer. It served as the post office until 1958. The building was in the ownership of Winifred Mayer Thorpe at the time of her death. This photograph is not dated.

as postmaster. The post office was originally situated in Mayer Station. Gradually, the post office outgrew the Mayers' home. By 1916, Mamie Mayer had taken over the responsibilities of postmaster from her mother, with assistance from Sadie. In 1917, a new post office was constructed next door to the Mayer Business Block on land owned by the Mayer family. The March 1918 issue of *Yavapai* magazine states that the post office "was erected by Miss M. Mayer and presented to the government." If, this was indeed the case, there must have been some provision that the building revert to the ownership of the Mayer family, as the building was owned by Winifred Mayer Thorpe at the time of her death. It was subsequently sold out of the Mayer family and into private ownership. By 1929, J. E. Harris was the postmaster. In the 1960s, this building was known as "Vaughn's Bargain Shop." This building still stands and looks much the same as it did when it was built. In 1958, another "new" post office was constructed on Main Street. The post office was constructed by Mr. Emil Mueller who donated the land and his services. A March 14, 1958, newspaper article in the Prescott *Courier* reported that "the historic town of Mayer, perched on the slopes of the Bradshaw Mountains, will dedicate a new post office Saturday—the first modern postal facility in the area since 1920. The present postmaster, Mrs. Clare Messard [sic], is the fourth since Sadie B. Mayer, wife of the founder of the town, was first commissioned in 1884." The postmistress was actually Mrs. Clare Lessard. Another "new" post office was constructed at the corner of Central Avenue and Main Street in the 1980s.

One of the amenities to come to an "up and coming" community was the establishment of churches. In an interview for the *Prescott Courier* in 1982, Irene Thompson McDonald mentioned that Mayer always had a community church and that she was a "good sized child before the Catholic church was here." Irene was born in 1905. The original Catholic Church must have been built around 1910 (one source says 1908) and is now a private home. At the present, Mayer has five churches, including the Community Christian Church, the Mayer Community Church, the First Baptist Church and the First Southern Baptist Church and St. Joseph's Catholic Church.

The Mayer library was established in 1969 by the "Library Committee," consisting of Mrs. Robert Combs and Mrs. Wesley Segner. Not wanting to wait for the completion of the construction of the Civic Center, which was to include a meeting space, kitchen and library, Mrs. Combs and Mrs. Segner converted a workshop in Mrs. Combs' home into a "cozy library" of 500 donated books. May 3, 1969, was to be the "grand opening" of the temporary library. Located on South Wicks Avenue, the current library is an up-to-date modern facility which serves the community of Mayer and the surrounding area.

Not every business was represented in the community directory. As mentioned earlier, the ranchers and farmers in the outlying areas were important to the commerce of Mayer, not only as customers, but as suppliers of goods. Ranchers sold or bartered their livestock for "store bought" merchandise. Leverette P. Nellis was a cattleman who had a meat market and store in Mayer, which, unfortunately, burned in the Mayer Owl fire of 1931. Farmers brought some of their produce, hay and grain to the local markets. Ranch and farm women also had their own enterprises. Eggs, butter, milk and cream, jams and jellies were common commodities to be sold or bartered in town. However, Claire Champie Cordes, who lived on a ranch with her husband Fred and

The Catholic Church, on South Jefferson Street, was built about 1910. It was one of the first churches in Mayer. Ish Davis donated the land and Joe and Sadie Mayer, who were Catholic, contributed funds to help build the church and a donated piano. It is now a private residence. Photograph by Rosema Promberger Minucci, 1941.

Built about 1970, the Mayer Library serves the Mayer area with a modern, up-to-date library. The library has a collection of historic photographs and documents which are under the management of life-long Mayerite Mona Bennett. Photograph by Nancy Burgess, 2011.

children at Turkey Creek (now the town of Cleator), wrote in *Ranch Trails and Short Tales* of her candy-making enterprise:

> We heard about a good milk cow for sale in Crown King. Fred was at camp so I decided to go get her. I rode Old Happy, my horse, up to Crown King, about twenty-five miles. I paid for the cow and started back home, down the old railroad tracks.... Happy was quick as lightning, ducking back and forth to keep Bossie in the right direction. My baby was about a year old, so I could hold her firmly in front of me on the saddle with one hand, and rein Happy with the other.... It took all day to get back home, but we made it.
> The cow was a blessing; we had milk, cream and butter. With the butter and cream, I was able to get into the candy business. Everyone bragged on it, so I decided it would sell. I sent a sample to Mrs. Wells' store at Mayer. The Swiss people who lived there appreciated the rich caramel candy and soon quite a business was built up, including divinity and cactus candy. Fred would always help me take the boxes to the train every week, and this went on for several months until old Bossie went dry and took a four-month rest. We had more than paid for her, and earned money we needed for Sears Roebuck clothes and other things. After her next calf, I was in business again... I made and sold candy for years. The cash came in handy during those lean years when we felt lucky to get three cents a pound for our yearling calves.

In an interview in June of 1981 for the *Prescott Courier,* Winifred Mayer Thorpe reminisced about Mayer in the "old days," the days when the town was experiencing its adolescence in a dusty swirl of mining activity. Miners, Indians and Chinese tracked through the streets, stores and bars of the town and entrepreneurs flashed on and off the scene "like sunshine through the clouds." Those were heady days for Joe Mayer's town. Looking back, and comparing the Mayer of yesterday with the Mayer of 1981, the reporter wrote of Winnie's reminisces:

> They still sing in Mayer's churches. They still hold an occasional dance. The children still run down to the creek bed to play. But today's Mayer is not the same as the one of Winnifred Thorpe's childhood.
>
> Mobile homes dot the hillsides. Winnebagos and pickup trucks have replaced the horse and buggy. Most of the area's mines are shut down.
>
> "The buildings are larger and there are more of them. More people have moved in" Mrs. Thorpe says. "I don't know as many people as I used to."
>
> But when you boil the trappings of progress away, the folks in Mayer haven't changed all that much to Winifred Thorpe's way of thinking. "I don't see any difference," she says, "maybe just the way they dress."

8

The Mining Boom: Boom and Bust

"The Man Behind the Pick"

There has been all kinds of gush about the man who is behind.
And the man behind the cannon has been toasted, wined and dined.
There's the man behind the musket and the man behind the fence;
And the man behind his whiskers and the man behind his rents;
And the man behind the plow beam and the man behind the hoe;
And the man behind the ballot and the man behind the dough;
And the man behind the counter and the man behind the hill;
And the man behind the pestle and the man behind the pill;
And the man behind the jimmy and the man behind the bars;
And the Johnny that goes swooping on the stage behind the stars;
And the man behind the kisser and the man behind the fist;
And the girl behind the man behind the gun is on the list.
And the man behind the bottle—and when they were short of men,
There was some small rhymester warbled of the man behind the pen.
But they missed one honest fellow and I'm raising of a kick,
For they don't make any mention of the man behind the pick.

Up the rugged mountain-side a thousand feet he takes his way,
Or, as far into the darkness, from the cheering light of day;
He is shut out from the sunlight, in the glimmer of the lamps;
He is cut off from the sweet air, in the sickly fumes and damps;
He must toil in cramped positions, he must take his life in hand.
For he works in deadly peril that but few can understand.
But he does it all in silence and he seldom makes a kick,
Which is why I sing the praises of the man behind the pick.
He unlocks the bolted portals of the mountain, to the stores
Hid in nature's vast exchequer in her treasure house of ores.
He applies the key dynamic and the gates are backward rolled,
And the ancient locks are riven to their secret heart of gold;
Things of comfort and of beauty and of usefulness are mined
By this brave and quiet worker—he's friend of human kind,
Who though tramped down and underpaid toils on without a kick,
So I lift my hat in honor to the man behind the pick.
    —Bert A. Judd, Hesperus, Colorado; published in the *Denver Post* and republished in the *Prescott Prospect*, January 1902.

The history of mining in Arizona is long, complicated and many-faceted. Mining has helped to shape the spirit of Arizona. Native Americans, Spaniards, Mexicans and

Europeans have mined in Arizona and have helped to establish the culture and personality of the state. Ever since Coronado went in search of the "Seven Cities of Cibola" in 1540, as Grace M. Sparkes wrote in "Yavapai, the Land of Opportunity" in August 1917, "Legends of immense treasures buried in the deep recesses of the mountains and guarded by the fierce Apache, have made the name of Arizona a synonym of magic wealth, mystery and romance." Mining brought people from many places and of many cultures to Arizona in search of fortune: many failed; many succeeded; many failed and then succeeded; and many succeeded and then failed. The business of mining in Arizona is fraught with death, injury, claim-jumping, jealousy, mystery, romance, fraud, great financial gain and great financial loss, political influence and political ruin.

If it were not for the discovery of gold in what is now central Arizona in 1863, the creation of the Territory of Arizona might have been delayed for decades. Most thought that the land acquired by conquest from Mexico along with the Gadsden Purchase gave the United States a vast, un-tillable wasteland of rock and sand which was to be ignored. In *History of Arizona Volume I*, Thomas Edwin Farish wrote of the influence the discovery of gold had on the organization of the Territory of Arizona: "Reports concerning these discoveries of gold in Arizona which were, no doubt, greatly magnified and exaggerated the farther they were carried, probably induced Congress to organize the Territory of Arizona, as the Government, at that time, was much in need of gold."

The pioneer miners of Yavapai County, the old "Hassayampers," endured the hardships and dangers of Indians, weather, medical emergencies and lack of food, commodities and companionship. They placer mined, maybe did a little sluicing if water were available, carried their belongings on their backs or on the backs of their burros, camped where they thought the "diggings" looked good and, when they proved not to be, moved on to another prospect. If they found a little (or a lot) or gold, word soon got out. Some miners were able to keep the secret of their find for a time, but once they went to town for supplies, (which the miner paid for with gold dust or nuggets,) his secret was up. Then the miner had to "prove up" his or her claim and frequently either sold out for a "tidy sum" or became part of a

J. E. Addicks, "the man behind the pick," poses with his pick over his shoulder at his Mattie R Mine in Yavapai County, 1903. Addicks was from a well-to-do East Coast family and came to Arizona to travel and do a little mining "just for the adventure of coming."

8. The Mining Boom 137

J. E. Addicks in front of his cabin in Yavapai County. The photograph is from his personal album of his trip to the "wild, wild west" and is dated January 12, 1903. There was a "girl he left behind him," but he returned to Maryland and married her.

A fine, new home in the mining camp at the Copper Cobre Mine, near Crazy Basin, Yavapai County, A. T. The notation on the photograph reads, "One of the most beautiful places in the country on a moon light night."

An "Old Hassayamper" sits in the doorway of his "make do" dwelling in Yavapai County about 1890. His home is covered with odds and ends of metal; some is pressed tin, all recycled from other places and uses (Sharlot Hall Museum).

new camp which sprang up around him. In *Sharlot Herself*, Nancy Kirkpatrick Wright published Sharlot Hall's description of the "Old Hassayampers" from a speech given before the Arizona Pioneers' Association:

> Those old pioneers who lighted her (Arizona's) first camp fires and blazed the first trails across her deserts and mountains, leads us back in memory to the days when to the average Hassayamper home meant a log cabin under the pines, a fireplace full of pitch logs and rose-colored dreams of the girl he left behind him.
>
> The Hassayamper was a pioneer of pioneers (for the reason that they were nearly all from other territories). His religion was to be a man among his fellow men. There were no churches, courts, schools or towns in the country in his day—every man was his own doctor, lawyer, preacher, blacksmith and judge of good whiskey.
>
> It might be that in that old-time cabin home he had nothing to offer but a pot of Arizona strawberries—the brown beans—or some venison jerky, but you were welcome to what he had and it was the universal rule that a crowd of Hassayampers would hang a man quicker for refusing to eat beans than for stealing a mule.
>
> There were two things from which you could not separate the true Hassayamper—his sack of brown beans and his burro—and here I move that when the new home is built and occupied, a pot of brown beans be kept always boiling on the fireplace and that the back yard be fenced off to accommodate the gray and saddle-scarred remnant of the old pack trains—the faithful burros who in their humble way were as true Hassayampers as their masters and had no small part in making the Arizona of today.

8. The Mining Boom

Without their "canaries of the desert," the miners of the southwest would have had a very difficult time of transporting their equipment and personal possessions, let alone their ore. These men are leading a small pack train of four burros packing ore from the Arizona Redlands Copper Company mine near Mayer in 1928. The caption reads, "Bringing down copper for Uncle Sam's pennies." Photograph by Bate.

In the 1890s, the demonetization of silver spurred an increased interest in the search for and development of new gold mines. Arizona became one of the regions which was most attractive to the mining speculator of the time. Although Arizona had a reputation for the production of only surface gold, by the 1890s it had been proven that Arizona had vast resources of "deep workings" of gold and other precious metals. In an article in the January 5, 1896, issue of the *Arizona Weekly-Journal-Miner,* the editor wrote of a field trip he had recently taken into the gold country of the Bradshaws. He wrote: "There is perhaps no section of Yavapai that can compete with the Bradshaws, Richenbar, Big Bug, Chaparral and the Agua Fria in the amount of smelting ore that could be brought

An unidentified Yavapai County mine photographed by J. E. Addicks in 1903. This is a typical small mine which could be worked by one or two men and a couple of burros. Abandoned mine tunnels and shafts dot the landscape of Yavapai County and can be very dangerous.

This hardrock mine crew in an unidentified mine are underground in the 1870s. Their only light is normally provided by candles, although these appear to be phosphorous lights, probably for the benefit of the photographer. The Miners' candlestick was driven into any handy crack in the rock or the timbering supporting the tunnel and held a single candle (Sharlot Hall Museum).

Placer mining involved a lot more than just panning for those elusive gold flakes at the bottom of the pan. Here, a 1920s Yavapai County placer miner uses a rocker, hoping to find "color" (Sharlot Hall Museum).

to a smelter. In this case, all roads lead to Mayer's, downhill from the mines in all directions and that point would be central for reduction works, as well as for the terminus of a branch railroad line."

But, in spite of the editor's promotion of the mechanized future of mining in the Big Bug District, he also wrote of his meeting with an old placer miner, which gives a very good idea of what placer mining by a well experienced individual miner was all about in 1896:

> Along the creek we met a congenial old miner rocking out golden flakes from the bed of the stream. He called his location the Klondike, and told us to wait a few moments and he would tell us how much his wealth had increased. He pulled the gravel into the rocker a pan full at a

8. The Mining Boom

As equipment and technology became available to the mines of Yavapai County, the mining companies developed stamps, mills and smelters. The buildings and the layout of the complex were always changing as advancements were made or problems came up. This is the Crowned King Mill at Crown King, circa 1900.

time and from an improvised bucket dipped water from the newly made trench, poured it into the rockaway and the oscillation of the crude machine shaking the finer particles down into a screen below. The reserve was then placed in a pan, taken to the stream, and by a process at which old miners are expert, the gravel was washed out of the pan, leaving on the bottom a neat saving of gold. Drying it in the sun, the gold was taken from the pan, placed in the hand and cautiously transferred to his brass purse in the shape of a cartridge to keep company with more of its kind.

By 1867, mining companies had been formed and the much sought after Eastern capital had started to flow into the Territory. Eastern capitalists were always eager for a new "get rich quick" investment and often were of the opinion that they were dealing with uneducated and unsophisticated "Hassayampers" of which they could take great advantage. But these mining men (and a few women) were often far more knowledgeable than the financiers suspected. Although they tended in the early years to be suspicious of the science of mining and the mining engineers who came to set up the mining companies, those mining companies who brought in heavy equipment and had the resources to build mills, smelters, rail systems to transport the raw ore, and the connections to make the mines pay, gradually began to come to the forefront in the mining history of Arizona, Yavapai County, the Bradshaw Mountains and the Big Bug District. One of those companies was the Big Bug Mining and Milling Company with the "Big Bug" mine, located about seven miles from Poland. An invoice dated May 14, 1867, to the Big Bug Mining and Milling Company provides a list of supplies purchased for the mine and gives the prices:

100 lbs. flour @$35 per cwt	$350.00
2 sacks bacon @$2.10 per lb.	105.00
30 lbs. coffee @60¢ per lb.	18.00
2 5-lb. cans baking powder	5.00
1 sack sugar, 120 lbs. @40¢	48.00
1 sack beans, 100 lbs. @40¢	40.00
25 lbs. dried apples @75¢	18.75
3 gallons syrup	12.00
5 lbs. tea @$2 per lb.	10.00
2 hams 30 lbs. @75¢	22.50
20 lbs. salt @35¢	7.00
1 box pepper	2.00
20 lbs. dried peaches	20.00
2 shovels	7.00
4 picks	16.00
4 pick handles	5.00
3 joints 6-inch stove pipe	6.00
One half dozen tin plates	3.00
Freight on com. 1051 lbs.	26.25
Total	$721.50

A belt of mineralization cuts across Arizona from the southeast (Bisbee) to the northwest (Page). As Grace M. Sparkes eloquently wrote in 1917, "Yavapai County is a region of mountains and valleys with a mean elevation of one mile. Crossing it diagonally is that great mineral belt, which signalizes its entrance to northern Arizona by the golden outburst of Mohave County and leaves as a parting gift, the world famous copper mines in and near Bisbee." Lying in the middle of this mineral belt is Yavapai County, the Big Bug Mining District and Mayer.

Discoveries of gold in what would become Yavapai County were announced by two separate exploration parties in 1862 and 1863. The Weaver Party, guided by Pauline Weaver, who had trapped extensively in Arizona in the 1840s and 1850s, located placers while camped at the base of a hill. The Rich Hill placer along with discoveries along Weaver and Antelope Creeks produced $500,000 in gold in five years. In 1861, the Walker Party was organized in California for the purpose of mining exploration to the east—in what would later become Arizona Territory. After several attempts and an arduous and dangerous trip, the party arrived in the early spring of 1863 at what is now the City of Prescott in Yavapai County. The bill establishing the Arizona Territory had already been passed by United States Congress in February 1863 while the Walker Party was en route. They camped about six miles south of Prescott and began placer mining along Granite Creek and other creeks adjacent to the area. The placers were extensive, including Lynx Creek, Turkey Creek and Big Bug Creek. On May 19, 1863, the Walker Party organized the Pioneer Placer Mining District. Each placer

8. The Mining Boom

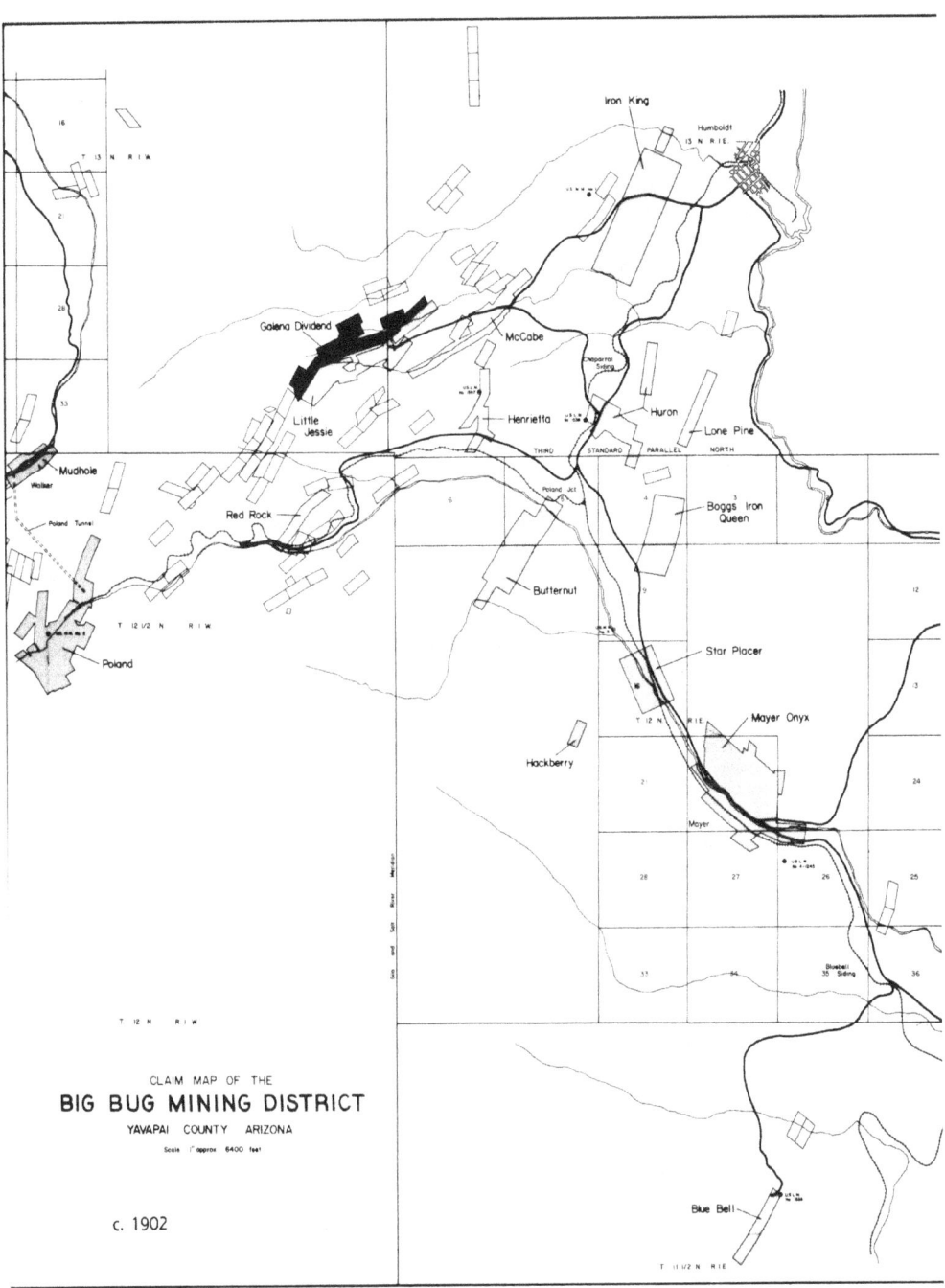

This 1902 era Claim Map of the Big Bug Mining District shows the P & E Railroad, including the Poland Branch of the Bradshaw Mountain Railroad, as well as the communities of Poland, Poland Junction, Huron, Humboldt and Mayer. Lynx Creek (upper left) and Big Bug Creek (center left to lower right) and the Black Canyon Road are also included. Mines shown on the map include the Poland (as well as the location of the Poland Tunnel from Poland to Walker), Henrietta, Boggs, Iron Queen, Hackberry, Star Placer, Butternut, Mayer Onyx and the Blue Bell.

claim was to be 300 feet long and 150 feet wide. Jack London captured the essence of every miner's dream when he wrote in *Call of the Wild*, "a shallow placer in a broad valley where the gold showed like yellow butter across the bottom of the washing pan."

The first finding of gold along the creek that would be named "Big Bug" was by the Walker Party. As Alvina N. Potter wrote in *Many Lives of the Lynx*, "The prospectors found another creek that ran east and it was not far from their woods. They found gold there, too, in paying quantities. While there, one of the men brought into camp, between two sticks, the largest bug they had ever seen. So that creek received the name 'Big Bug.'" Potter wrote further that "there were claims located on Big Bug Creek in June, 1863. Of the first thirteen, all were claimed by members of the original Walker Party except No. 2.... Many claims were located and recorded. At first these were placer claims in the creeks and the gulches. Soon the men discovered minerals in the lodes and ledges in the hills so lode claims were also located." In 1865, the *Arizona Miner* reported that "new diggings have been discovered at Big Bug which pay well, and a nugget, over $13, was found last week. The gulch is pretty long and has a good stream of water at present and affords plenty of employment. The ledge, known as 'Big Bug,' is turning out most satisfactorily." This was the start of the modern mining industry in Yavapai County and, in particular, the Big Bug Mining District.

In Eldred Wilson's *Gold Placers and Placering in Arizona*, a "Suggested List of Equipment for Prospecting in the Southwest," by Charles H. Johnson, was republished. The list includes the following prospecting tools: "Hammer (one single jack about four pound weight), shovel (round point, long handle), miner's pick, prospector's pick, moils (two or three), gold pan, horn spoon, small mortar and pestle, magnifying glass, blow-pipe outfit, determinative tables, sample sacks (some use double paper sacks), compass, maps (topographic and geologic). If the prospector expects to make a permanent camp and some hard rock mining: powder, caps, fuse, hand steel, spoon (to clean drill holes), tamping stick and blacksmithing equipment and tools."

There were more than 70 mining districts in Yavapai County, including the Big Bug District. The Big Bug Mining District was apparently formed in 1867. It was located in south-central Yavapai County in the general vicinity of Big Bug Creek and Mayer, Poland, McCabe and Humboldt. According to *Gold Placers and Placering in Arizona*, "Gold was discovered within the big Bug District in the late sixties, but the greatest activity in placer mining there was during the eighties of the past century (1880s). Considerable sluicing, rocking, and panning have gone on, especially in upper Big Bug Creek as far down as Mayer, and in Chaparral and other gulches near McCabe.... One of the largest nuggets found in the Big Bug region contained about $500 worth of gold figured at $29.67 per fine ounce." In a publication by the Prescott Mining Exchange in 1897 entitled "Prescott—The Mining Center of the Southwest," it was reported that "the principal industry is mining in which probably three-fourths of the population are directly or indirectly engaged. The products are gold, copper, silver and lead.... There are now in the county thirty-two gold mills, 390 stamps, and several plants are now in course of construction.... Among other valuable products of the county, are the onyx beds on the Big Bug situate 25 miles southeast of Prescott. These beds cover an area of about 200 acres, and average about 10 feet deep."

There were literally hundreds of mining claims in the Big Bug District. The naming

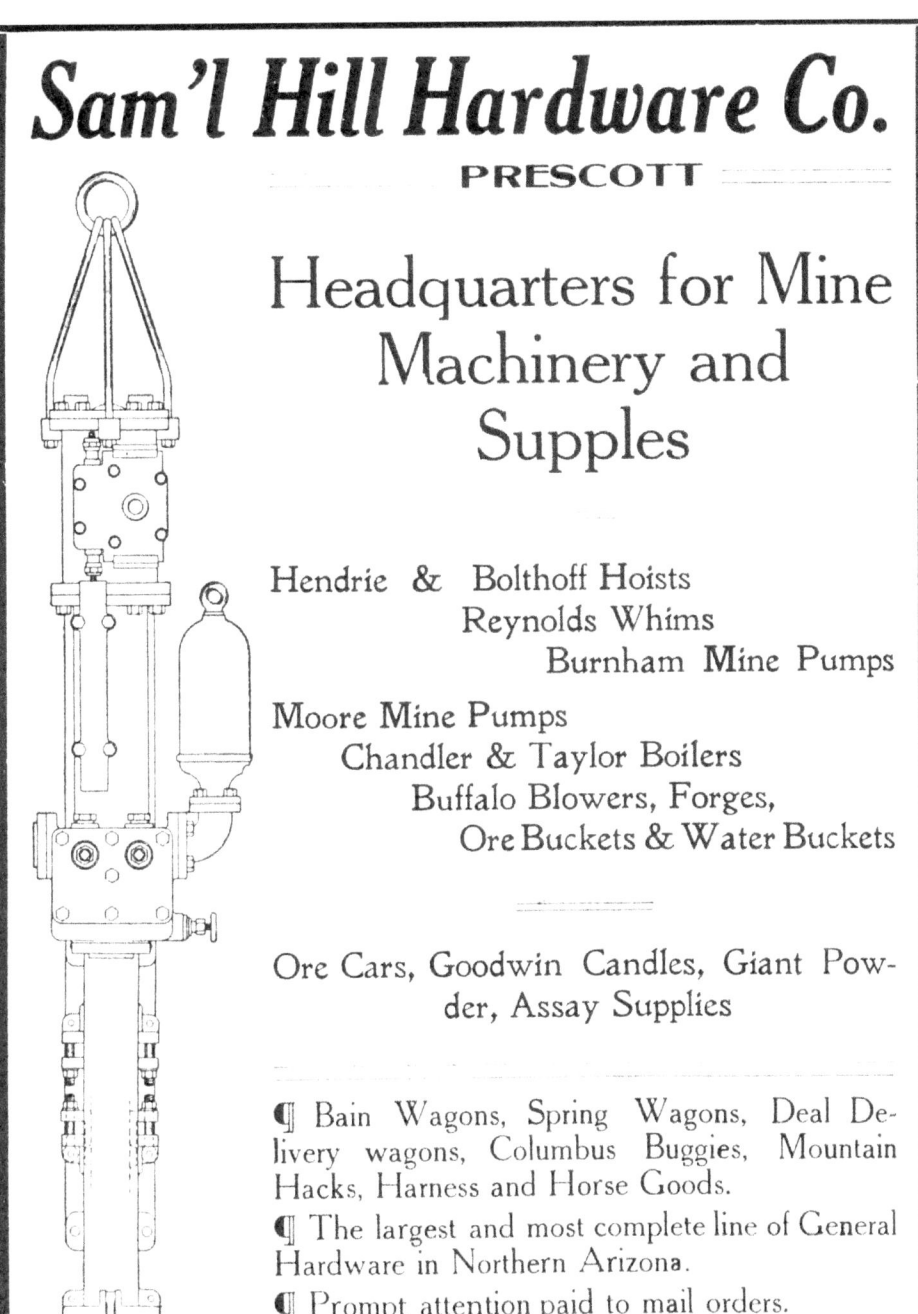

The Sam'l Hill Hardware Company and Arizona Mining Supply in Prescott were the major suppliers of mining equipment in western Yavapai County. Access to the railroad starting in 1886 gave Prescott businesses an advantage over the small communities that were not accessible by railroad until much later. Moore Mine Pumps, Chandler and Taylor boilers, Buffalo blowers, forges, ore buckets, water buckets, ore cars, Goodwin candles, Giant powder and assay supplies were included in this 1917 advertisement for mining supplies.

of mines was an art unto itself. Yavapai County is a treasure trove of creative names for mines. Some were named for the minerals found there, such as the "Iron King," "Gold King" and "Copper Queen"; some were named for places, such as the "Baltimore" or the "California"; some were named for people, such as the "Joe Walker" and the "Mabel," "Maggie" and "Ida May"; some were named for events, such as the "Accidental" or the "Bunker Hill"; some were named for the animals found there, such as the "Badger" and the "Rats Nest"; some were named for the plants in the area, such as the "Juniper," "Live Oak," "Cottonwood" and "Golden Lillie"; some were named for the frustrations of the miner, such as the "Swindler," "Slim Chance," "Hard Times" and the "Tempest"; and some were named for the hopes and dreams of a jackpot, such as the "New Hope," "Lucky Jack," "Last Chance," "First Chance," "Dividend," "Urika," "Hidden Treasure" and the "Benefit Trust."

In the July 1939 issue of *See Scenic Southwest*, an article about Mayer mentioned Mayer's role in the mining industry in the days before the railroad came to town in 1898: "So past the store, saloon and hotel of Joe Mayer rumbled the heavy ore wagons with their great long strings of oxen or mules, and up to his door rattled the stage coach on its way to and from Phoenix, and through his door clanked the cowboy and thumped the miner and Joe Mayer prospered." Joe Mayer wasn't a miner, but like almost anyone in Yavapai County in the 19th century who had a few dollars in his pocket, he was a mine investor and speculator and, at one time, even managed mines. A March 16, 1906, article in the *Arizona Republican* regarding Joe Mayer stated: "Joe

A ten-hitch team is hauling mining equipment through Mayer en route to a smelter. On the back of the photograph it says "delivering an express package to the smelter, Mayer, Arizona" (courtesy Arizona Historical Society/Tucson, no. 49379).

8. The Mining Boom   147

This circa 1910 postcard of Mayer shows a miner with his two burros in the foreground, perhaps one of the men Joe Mayer would have "grubstaked." The photograph was taken facing northwest toward downtown Mayer. The White House Hotel is on the right with the bunkhouse in the back. Railroad tracks run between the bunkhouse and railroad warehouses. Joe Mayer's brick warehouse is in the center. The P & E Depot is on the left edge of the image. The Mayer Hotel and the Mayer Business Block were off the edge of the photograph in the upper left. Real photograph postcard.

Mayer, the man who owns the town of Mayer, is always mining and successfully, too, I am glad to say, but mentioning all of his properties would take too much space. In the old Moscow Mine he tells he has opened up exceedingly rich bodies of good ore and I know of other prospects of his that look good to me. He has a mine in the old Gillespie property and it will be heard of someday and numerous interests in the Black Canyon and Turkey Creek sections." Joe Mayer's favorite mining adventure usually involved grubstaking someone who was down on his luck but knew the next strike would be his bonanza. This is how Joe Mayer lost a lot of money, but also how he and his wife and children sometimes ended up owning mining claims in Yavapai County. In fact, the family located and owned dozens of mining claims, both outright and in partnership with others. Some of the Mayer family mining claims included the "Cypress," "Hobgoblin," "Jickey No. 2," "Martie," "Winneford," "Gillespie," "Miner," "Yankee Girl" and the "French Lily," many of which were passed down from Joe and Sadie to Mamie, Martie, Wilbur and Winifred. Both the Mayer Station and the Mayer Business Block sat on top of mining claims filed by Joe Mayer when he first settled at Big Bug Station.

Although there were hundreds of mining claims and mills, numerous companies and smelters, railroads and businesses associated with mining in the Big Bug District,

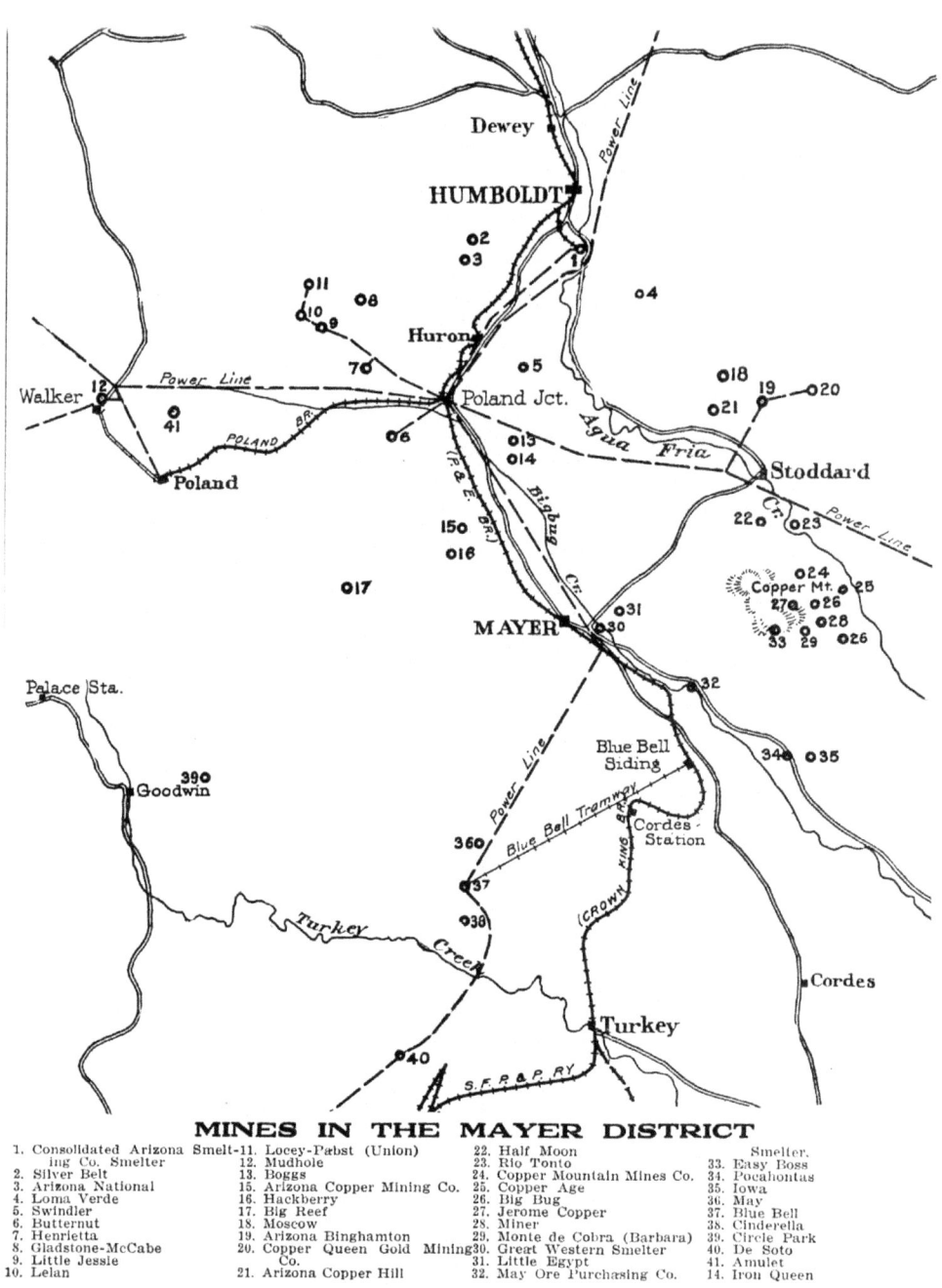

MINES IN THE MAYER DISTRICT

1. Consolidated Arizona Smelting Co. Smelter
2. Silver Belt
3. Arizona National
4. Loma Verde
5. Swindler
6. Butternut
7. Henrietta
8. Gladstone-McCabe
9. Little Jessie
10. Lelan
11. Locey-Pabst (Union)
12. Mudhole
13. Boggs
15. Arizona Copper Mining Co.
16. Hackberry
17. Big Reef
18. Moscow
19. Arizona Binghamton
20. Copper Queen Gold Mining Co.
21. Arizona Copper Hill
22. Half Moon
23. Rio Tonto
24. Copper Mountain Mines Co.
25. Copper Age
26. Big Bug
27. Jerome Copper
28. Miner
29. Monte de Cobra (Barbara)
30. Great Western Smelter
31. Little Egypt
32. May Ore Purchasing Co. Smelter.
33. Easy Boss
34. Pocahontas
35. Iowa
36. May
37. Blue Bell
38. Cinderella
39. Circle Park
40. De Soto
41. Amulet
14. Iron Queen

A less detailed and easy to read map from the March 1918 issue of Yavapai *magazine shows Mayer in the center of the map, the locations of Dewey, Humboldt, Huron, Poland Junction, Poland and Mayer. Number 13 on the map is the location of the Boggs Mine; number 14 is the Iron Queen Mine; number 16 is the Hackberry Mine; number 30 is the Great Western Smelter; and number 32 is the Mayer Ore Purchasing Company Smelter.*

8. The Mining Boom 149

A photograph taken in the fall of 1899 of the Mayer Onyx Quarry shows the cable arrangements set up to move the large blocks of stone to be transported to the P & E for shipment. Photograph by Putnam.

there were a few which had a particular association with Joe Mayer and his town. The first of these is the Mayer Onyx Mine, which was located in 1889 by Joe Mayer, Frank Murphy, Jesse Davis and Al McCann. The onyx beds covered an area of about 200 acres. In 1890, William O. "Buckey" O'Neill (of Rough Rider fame) became a partner in the onyx claims along with Joe Mayer. O'Neill paid $150 for a one-third interest in the mine. In 1893, he and his partners purportedly sold their interests in the mine for $200,000. The onyx claims were bought and sold many times, and with changes in ownership, stops and starts in production, legal troubles, changes in the demand for the product and highway construction right through the middle of the onyx beds, it is miraculous that they are still in production today. The Mayer onyx is actually an onyx-marble, a crystalline form of lime carbonate. It comes in two varieties, one with layers of brown, red, green and white, which is called "Grand Canyon Onyx," and the second with black, grey, dark brown and white layers, which is called "Black Canyon Onyx." A report in the *Arizona Weekly Journal-Miner* in October of 1896 described the quarries, stating that "The onyx lays in a saucer shape, and the rim has been worked all around, showing the dip to the center." In 1897, the owners of the mine engaged George C. Underhill and his son, from Vermont, to operate the quarry. Underhill brought in machinery to operate the mine, and under his management, the Arizona

A circa 1915 postcard of the Mayer Onyx Quarry shows the striations of the stone. This is a Heil Drugstore (Prescott) postcard. Photograph by Erwin Baer.

One of the several owners of the Mayer onyx quarry was American Onyx Products, Inc. Based on the telephone number, they could have been in business in the late 1940s up through the early 1960s.

Onyx Company sparked a boom in the economy of Mayer. According to the *Arizona Weekly Journal-Miner* in January of 1898, the arrival of the railroad to Mayer was greatly anticipated by the owners of the onyx mine. The company, the "International Onyx and Marble Company," had at the time hundreds of tons of stone cut and ready to ship, including an order of $100,000 worth of stone for a "big structure for R. G. Dunn & Co. in New York City." In February 1901, the *Prospect* reported that "Dan Bowen is finishing up the assessment work on his onyx claims. He opened up some very highly colored, fine grained, sea green onyx this week." A promotional brochure published in 1907 by the Yavapai Commercial Club described the onyx deposits in Mayer as being "extremely variable in texture and color. Much of it is of a general coloring, and ranges from whites and creams to pale greens; pink and salmon tones to dark iron browns. The beautiful bandings and shadings make this onyx especially valuable for decorative work." In 1908, the onyx quarries again had new owners, and the *Courier* reported that they had "about 25 slabs of immense size on the dump and are shipping a car of onyx a day to various points. They are now in daily expectation of hearing from the carload they shipped to London." In fact, Mayer onyx was used extensively in the automobile industry in the United States, as seven manufacturers were using Mayer onyx as decorative detailing in their automobiles. In 1926, more than 1,000,000 pounds of onyx were shipped. The quarries were shut down during the Depression and were idle for many years. In the 1960s, Jack Givens ran the business as "Givens Stone Yard." In 2008, the business was operated by "Stoneworld." Their marketing director states that it is one of the largest onyx quarries in the United States and that there is enough onyx at Mayer for the "next 200 years." The onyx can easily be seen today by the thousands of motorists who drive Highway 69 through Mayer, and also from the playground of the Red Brick Schoolhouse in Mayer. Recently, the Mayer Onyx Mine was again for sale.

Once Frank Murphy established the Poland Mining Company in 1900, a flurry of activity began to develop the mine. The Poland Mine was located just a few hundred feet above the townsite of Poland. The ore was processed by this 20-stamp mill located between the mine and the railroad spur. This photograph of the mill is dated 1902.

The Poland Branch of the Bradshaw Mountain Railroad left the P & E at Poland Junction south of Huron and headed west along Big Bug Creek to the Poland-Hamilton Mine, which was acquired by Frank Murphy in 1891, and the town of Poland. In 1901, the Poland-Hamilton group of mines consisted of six claims: Yukon, Poland, Hamilton, Chalcedony, Jasper and Wapakoneta. In all, the company owned 26 claims between Big Bug Creek and Lynx Creek. Regarding the Poland-Hamilton Mine, an article in the June, 29, 1901, issue of the *Prospect* stated:

> The isolated location of the Poland-Hamilton precluded any attempt towards making it a profitable producer. Nevertheless, it was a pet of the company and development steadily progressed on the lines of the old adage "All things come to him who waits." Now that time has come.
>
> As the necessary development has been carried out which opens up this property, exposing its vast sources of wealth; as the necessary facilities for the marketing of the ores by the extension of the railroad to the mill site has been secured; the conditions for the successful operations of the company's mines have been realized; steps have now been taken to actively operate these mines on a huge scale which will give it a high rank among the dividend payers of the territory.

Frank Murphy, who was always interested in developing both his mining and railroad interests in Arizona, desired to have a branch line to his and E. B. Gage's Poland-Hamilton mining claims. The Bradshaw Mountain Railroad was incorporated in February 1901 with $1,050,000 in capital stock. The survey for the Poland Branch line was completed in early 1901. Construction on the Poland Branch of the Bradshaw Mountain Railroad started on September 14, 1901, while negotiations were underway to sell Murphy's Santa Fe, Prescott and Phoenix Railway to the Santa Fe. The 7.79 mile-long branch line opened on May 11, 1902, with a special excursion run from Prescott. Stations along the line included Torres (Henrietta Station), Eugenie, Providence (established by the railroad to serve the Sterling Belcher and Red Rock mines), Block (a side track named for Prescott merchant Ed Block, who also owned the Merchants Mining Company) and Poland. Mines in the area included the Henrietta Mine, which at one time was

Six people and a dog are on the platform of the small Poland Depot. The depot looked down on the turntable. The Poland townsite was dominated by the structures of the Poland Branch of the Bradshaw Mountain Railway. Real photograph postcard.

The town of Poland started to take shape in 1901. The post office was established November 16, 1901. Many of the merchants from Providence bought property in Poland Canyon in anticipation of the arrival of the railroad. Homes, businesses and warehouses dotted the landscape up and down the hillsides.

managed by Joe Mayer, the Red Rock Mine, the Eugenie Mine and the Poland Extension Gold Mining and Milling Company. It took 45 minutes to traverse the eight miles of the Bradshaw Mountain Railroad line while the Brooks 2-8-0 locomotives and their cars chugged through the only tunnel on the line, groaned on the short curves and struggled up the 4 percent grades.

The mine and the soon-to-be community of Poland were named for a pioneer of the Big Bug District, Davis Robert Poland. Originally just a camp at the turn of the century, Poland was built in a rugged and narrow canyon. The town climbed up and down the steep canyon sides, and Big Bug Creek ran through the canyon. The mine and mill were located on the slope a few hundred feet above and northeast of the townsite of Poland. Speculation about the future arrival of the railroad resulted in a boom in building activity evidenced by a new hotel, several stores and "innumerable" saloons. By the time the railroad arrived in May 1902, the town of Poland was bursting with more new construction and new people. The Poland Mill, after a delay caused by a lack of water, reopened in December 1902 and was soon running 24-hour shifts. The Walker-Poland Tunnel, which began at the west end of the Poland Mill and ran 8,071 feet through the mountain to the mines of the Walker Mining District, gave

A 1902 view of the Poland Mill shows the waste pile from the mine and the lift to take the ore down to the mill. The entrance to Walker-Poland Tunnel would be drilled starting just above the mill to the right.

access to the railroad for the mines along Lynx Creek and provided ore from the Walker area mines to the Poland Mill for processing. It took three years to build and was an engineering feat. On May 8, 1904, the crews drilling from the Walker side broke through to the tunnel from the Poland side. They were less than six inches off alignment. Initially, burros were used to pull the ore cars along the track through the tunnel. In about 1910, electric power provided by the Arizona Power Company from its Poland Junction Substation to Poland gave Frank Murphy the opportunity to install an electric railway in the tunnel. The tunnel, at 8 by 12 feet, accommodated the double-track mine-gauge railway. Once the school in Poland closed down, the children walked through the tunnel to Walker to go to school.

The Poland Mine and the town of Poland had a major impact on the Big Bug Mining District. One of the best-producing mines in the area, plus the business of the Poland Mill, made the enterprise a successful one—at least for awhile. In 1905,

8. *The Mining Boom*

J. E. Addicks took this photograph of the Little Jessie Mill near McCabe in 1903. An October 1895 article in the Arizona Weekly Journal-Miner states that "at a depth of 550 feet the Little Jessie Mine has taken on quite a supply of water and John S. Jones has completed a pipeline to the mill for utilizing it."

Frank Murphy's Development Company of America had the old mill buildings demolished and built a new 20-stamp mill and concentrator. In 1908, the company mine produced gold, silver and copper, but the major production was lead. By 1911, though, most of the mines had played out and the traffic on the Poland Branch slowed to a trickle, and to add to the problems, the ten-year tax exemption for the railroad had expired. The Poland Mine closed in 1912. World War I brought renewed activity when ore prices rose, but by 1919, mining activity in the area had ceased. Frank Murphy died after a year-long illness on June 23, 1917. In November 1919, the Poland Mine and Mill buildings were all demolished, with the material being salvaged for other uses. On July 16, 1920, the Poland Branch of the Bradshaw Mountain Railway line was officially abandoned. Today, the Poland Tunnel is a favorite location for geocaching.

The history of mining in the Big Bug District cannot be properly documented without mention of Theodore Warner Boggs, the first settler in the area of what was later called "Big Bug," and the developer ot the Boggs and Hackberry mines. In 1863, Theodore Boggs came to Arizona from California and established a ranch in a shady grove on Big Bug Creek, much as Joe Mayer did almost 20 years later. Boggs had a reputation for treating all guests royally. He prospected over a 30-year period, locating

Theodore W. Boggs prospected over a 30-year period and located numerous mines, including the Little Jessie and the Hackberry. During the 1890s, the Hackberry Mining Company was controlled by the Phelps Dodge Company. The "baby" gauge Hackberry & Boggs Railroad carried ore from the Hackberry Mine to the smelter at Curtiss (Arizona City). There is one engineer in the cab and another man standing on one of the three ore cars. An 1898 classified advertisement requested "woodchoppers, at or near Boggs Mine. Price per cord, $1.50." Note the immense piles of wood in the upper right of the photograph. Photograph by E. M. Jennings.

several important mines including the "Little Jessie," the "Butternut" and the famed "Hackberry" and "Boggs" mines. An article in the *Phoenix Gazette* in 1893 stated that Theodore Boggs "did more to turn the light on the treasures of the Big Bug District" than anyone else.

Big Bug was a settlement just about two miles from what would later become Poland Junction on the Poland Branch of the Bradshaw Mountain Railway line. At the "gateway" to the town was the Annie Mine. There was a post office at Big Bug from 1879 to 1910. The *Arizona Weekly-Journal-Miner* of January 5, 1898, ran a lengthy article concerning a visit the reporter had recently made to the McCabe and Jessie Mines. On the trip back to Prescott, the reporter stopped off at Big Bug to visit pioneer Theodore Boggs, stating: "Theodore Boggs is a pioneer citizen of the Big Bug country, having located in that section in 1863. There are few, if any, valuable mining properties in that territory in which he had not been interested. The Boggs Mine, two miles east of his place, is well known to the mining fraternity as the property once operated by the Commercial Mining Company, a smelter being run in its conjunction." On June

Opposite: *This map clearly shows the route of the Hackberry & Boggs Railroad from both the Hackberry and Boggs mines to Arizona City and the Boggs Smelter where the ore could be transferred from the narrow gauge cars to the standard gauge P & E cars. Map courtesy of David Myrick,* **Santa Fe to Phoenix, Railroads of Arizona Volume 5.**

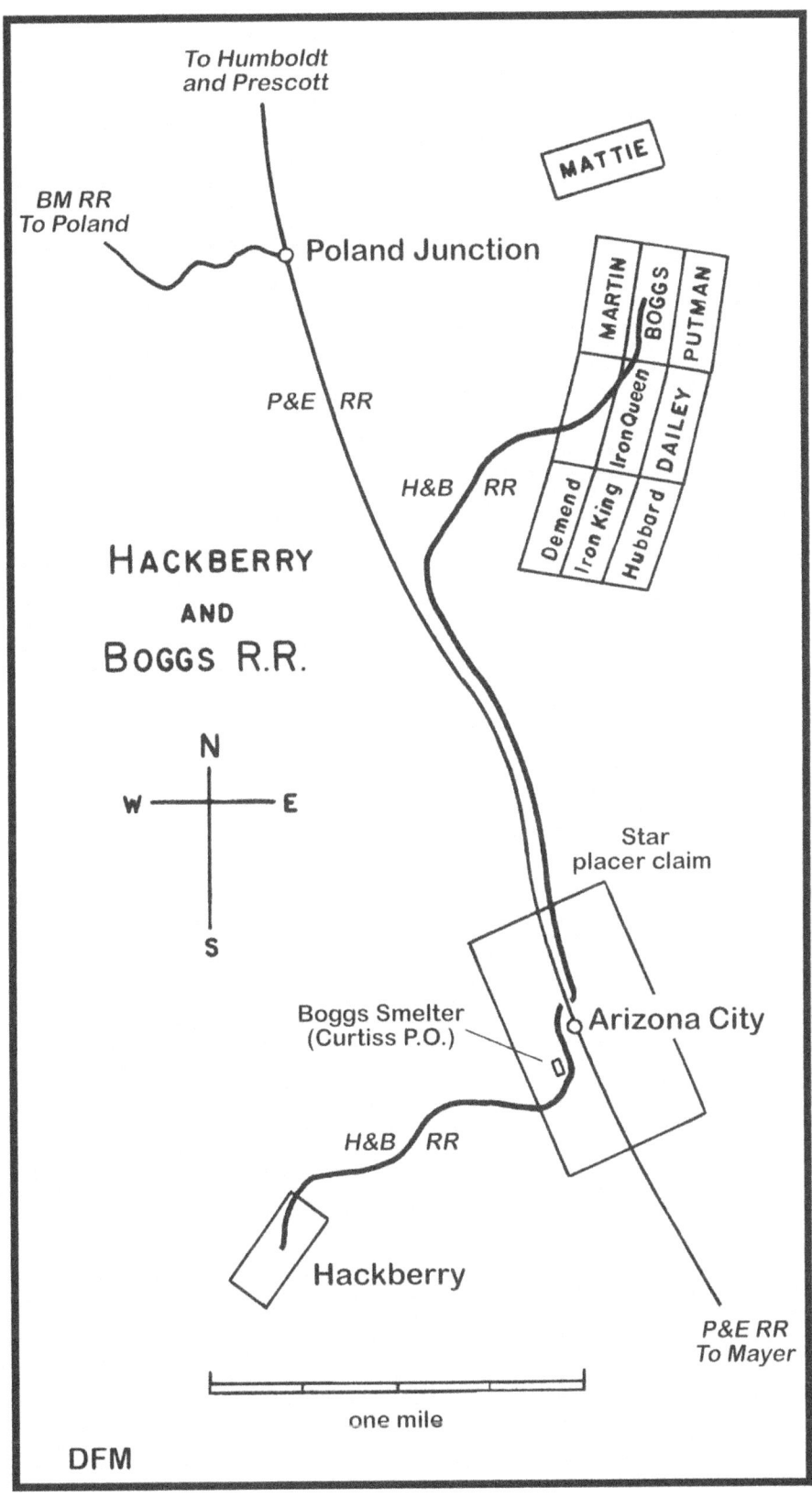

The General Store at Curtiss (later called Arizona City), which was named for the superintendent of the nearby mine, served the area miners and the operators of a small mill in the 1880s in the area along Big Bug Creek (Sharlot Hall Museum).

Here, Engine No. 1 of the Hackberry & Boggs Railroad carries some well-dressed folks across a trestle about 1891 (Sharlot Hall Museum).

29, 1881, Boggs located his Hackberry Mine. Located a few miles west of Big Bug Creek, the mine was not activated until 1888 when the Big Bug Reduction and Development Company was formed with seven mining claims under its corporate hat, including the Boggs and the Hackberry.

The Boggs mine was located east of Big Bug Creek, southeast of Poland Junction. The distance between the two mines was about four miles. In the spring of 1889, Boggs and his associates sold their Big Bug Reduction and Development Company to Phelps Dodge. On September 4, 1889, Phelps Dodge incorporated the Hackberry Mining Company. During the 1890s, an affiliate of Phelps Dodge, the Commercial Mining Company owned and managed the Hackberry Mine. Under the watchful eye of Dr. James Douglas, improvements were made. On a visit to the Boggs-Hackberry operation in September of 1890, Douglas authorized the surveying of a 20-inch gauge railroad to consist of two lines—one from the Boggs Mine and one from the Hackberry Mine—which were to meet at Arizona City, just north and east of the Boggs Smelter. As David Myrick wrote of the Hackberry and Boggs Railroad in *Santa Fe to Phoenix, Railroads of Arizona Volume 5*, "Grading followed and by the end of November, a diminutive locomotive, an 0-4-0 built by H. K. Porter (#1201), 'No. 1' of the Hackberry RR or Hackberry and Boggs Railroad (the informal name of the railroad varied slightly as the road lacked a home in a separate corporation) arrived in Prescott and was forwarded to Big Bug Creek in an ordinary freight wagon." The Boggs Smelter, also called the Big Bug Smelter, was completed, along with the Hackberry and Boggs Railroad, in the spring of 1891 and the train made two trips a day bringing ore to the smelter. The smelter was an impressive operation for its time. However, the Prescott and Eastern had not yet been constructed, so the ore from the smelter still had to be

Located very close to Big Bug Creek, the Boggs Smelter (also called the "Big Bug Smelter"), which appears to be going "full blast" in this 1898 photograph, processed ore from the Boggs and Hackberry mines. Today, there is nothing left of this operation but a slag pile. Photograph by E. M. Jennings.

Several years after purchasing Porter Engine No. 1, the Hackberry railroad acquired a second Porter 0-4-0, shown here with ore cars in tow. The higher trestle in the rear allowed the transfer of ore from the narrow gauge to the standard gauge cars. Newspaper articles report that this engine was sold to Japan for scrap. Photograph is circa 1904 (Sharlot Hall Museum).

transported by ore wagon to Prescott to be shipped. By 1897, the railroad was no longer operating, and by the time the Star Placer Claim had been surveyed in July 1898, the tracks had been removed. Later, Dr. George A. Treadwell leased the former Commercial Mining Company properties at Big Bug. Near the old smelter, Treadwell constructed a new-style "hydro-carbon" smelter using fuel oil as its heat source. It was "blown-in" (the first firing) in December 1902. Meanwhile, Treadwell's son, Edwin D. Treadwell, decided to restore the then-abandoned Hackberry and Boggs Railroad. However, after linking the Hackberry to the new smelter with new rails, his plans were dashed when a summer storm sent torrents of water from Grapevine Canyon onto the Hackberry Railroad, and according to an article in the *Coconino Sun* on August 3, 1903, "resulted in washing out about a mile of the little narrow gauge railroad from the Big Bug Smelter to the Hackberry Mine. The grade, ties, bridges and rails, all went with the flood. The grade and ties were washed down the creek, never to be found, while the rails, all twisted, bent and broken, were scattered promiscuously along creek banks. Joe Mayer says the water was the highest in Big Bug he has seen since the flood of 1891." Two years later, Treadwell again restored the Hackberry portion of the baby-gauge railroad, this time with seven trestles. By the end of 1904, the "eastern" portion, to be called the "Hackberry and Iron Queen Railroad," of the former Hackberry and Boggs Railroad, was again in operation. The railroad connected the Hackberry, Boggs and Iron Queen mines, joining the P & E at Arizona City. Equipment included two locomotives and nine cars. Treadwell's "hydro-carbon" smelter was a "flat failure," although it was advertised as a "great success," and his plans for a well-proven "water-jacket" furnace never materialized. Treadwell decided to abandon all of his existing plants and replace them with a new facility just southeast of Mayer. Treadwell operated the Boggs and Hackberry mines in 1905 and 1906, and in 1907 leased them to the American Smelting Company. By 1909, they were idle, and in November 1910, the Commercial Mining Company foreclosed.

This was not, however, the last to be heard of Dr. George A. Treadwell. Having taken an interest in Mayer, Treadwell built a home there. He and his son, along with Joe Mayer and others, built a pipeline in 1902–03 from Crystal Springs to supply 400,000 gallons of water a day to Mayer. After Joe Mayer, George Scammel and George Treadwell platted the Mayer Townsite subdivision in 1904, development in the community increased. Soon after completing a spur track with a significant trestle from the P & E northeast of Mayer, Treadwell broke ground for a 250-ton smelter on July 25, 1904. In September 1904, the *Daily Journal-Miner* said of the Treadwell Smelter, "It is a gigantic and complicated enterprise mechanically and is being pushed to completion with the utmost care. The Treadwell Company, who are the owners of it, will probably utilize it solely on ore from their many claims." As David Myrick wrote in *Santa Fe to Phoenix, Railroads of Arizona Volume 5*, "Ore from the company mines came down from Arizona City on the baby gauge railroad (the Hackberry and Boggs) where it was dumped into the standard gauge gondolas for the final P & E haul over the new spur to the smelter. Blown in during the spring of 1905 the smelter proved unsuccessful. Treadwell explained that the Humboldt Smelter offered such a favorable contract, he saw no reason to operate the new smelter."

The diminutive Hackberry & Boggs Railroad engine No. 1 chugs along with three ore cars in tow. As John Sayre wrote about the Hackberry & Boggs Railroad in Ghost Railroads of Central Arizona, *"This railroad was a visual highlight of railroad travel on the Prescott & Eastern mainline, and passengers and crew alike strained for a view of the tiny engine at work."*

Of the George A. Treadwell Mining Company, the *Copper Handbook, Volume VI*, published in 1906, had the following to say:

> For seven years the George A. Treadwell Mining Company has been peddling its shares to widows, orphans and fools. For seven years its advertisements have promised impossibilities. For seven years this stock-jobbing corporation has paid out a large part of its income from stock-sales in advertising more shares for sale.... To list the lies uttered by this putrid corporation would require pages of this volume. The company has lied about its mines, its ores, its smelters, its furnaces and its earnings.... The advertisements of the company always carry the same burden—just a little more stock for sale—buy it quick and get rich.... The whole outfit should be jailed, for securing money under false pretenses.

From 1906 to 1915, the Treadwell Smelter lay idle on the hill above Mayer. Its buildings and machinery were vandalized and ransacked. In late 1916 or early 1917, the plant was acquired by the Great Western Smelter Company. The plans were for the smelter to process ores from two very rich mines in the area: the Henrietta Mine, eight miles west of Mayer, which was one of the oldest gold mines in the district; and the Butternut, which formerly was located by Theodore Boggs, and was a copper mine about three miles northwest of Mayer. Both properties had been purchased by the Big Ledge Mining Company in 1916 and were to be reopened after being "dewatered." Joe Mayer was involved with both mines, having managed the Henrietta at one time in the 1890s, which, unfortunately, ended up embroiling him in a lawsuit in 1895.

The George A. Treadwell Mining Company was formed on May 17, 1901. About a half-mile south of the Mayer Depot, the Treadwell Spur led to the Treadwell Smelter, built in 1905. It never was successful. In this 1917 photograph looking south to southeast, the Great Western Smelter, the successor to the Treadwell, shows the "hydro-carbon" smelter. Tenant houses are in the foreground (courtesy Arizona Historical Society/Tucson, no. 49376).

8. The Mining Boom   163

A closer view of the Great Western Smelter, probably about the time the Big Ledge Mining Company took over the property in 1917, shows the trestle leading from the spur to the smelter. The ore bins are at the end of the trestle. This enterprise was also a failure, and only the foundations remain.

The Mayer Ore Purchasing Plant was south of Mayer on the Black Canyon Road adjacent to Big Bug Creek. This Mayer Smelters Corporation Common Stock certificate was never issued (Arizona Collection, Arizona State University Libraries).

The Big Ledge Mining Company acquired the Great Western (Treadwell) Smelter in 1917. The old plant was stripped of everything which was not useful and rehabilitated for wartime use. Some of the structures were retained, but the hydrocarbon smelter was demolished and was replaced with a new smelter with a capacity of 600-tons per day. It was blown in on January 17, 1917. But the Big Ledge Development Company was in financial trouble at the very start. The new smokestack at the Great Western Smelter was begun in 1917 by the Weber Chimney Company of Kansas City. Although the Big Ledge Company tried to stop the construction of the smokestack, which was to be 129.5 feet high and would create a tremendous updraft and intense furnace heat, the construction company insisted on completing the job. Today, the smokestack is a landmark of Mayer, watching over the town from its hilltop just off Highway 69. The foundations and ruins of the Great Western Smelter, which closed in late 1917, are a popular spot for local explorers and picnickers.

Other processing plants dotted the hillsides around Mayer once the mining boom of the World War I era got underway. The Grey Eagle Reduction Plant on the Crown King Branch of the Bradshaw Mountain Railroad smelter was completed in early 1916. It promptly closed. Renamed the Mayer Custom Plant (Mayer Ore Purchasing Plant), the plant was reopened in the spring of 1918 by its original designer, Dr. H. A. Wagner.

The landmark Great Western Smelter Stack on the hill above Mayer, built in 1917, was photographed by Rosena Promberger Minucci in 1941.

It was intended to provide a "one stop shop" for small mine operators by purchasing raw ore of almost any type and in all quantities from a bag to a rail car and processing it on site or stockpiling it for shipment elsewhere. The plant closed in 1919 as World War I ended. The plant was dismantled and its equipment salvaged and sold.

Constructed just outside Mayer at about the same time as the Treadwell Smelter, the Rigby Reduction Works was built on 70 acres northwest of Mayer. The president and general manager of the company was Col. T. Johns Rigby. According to the *Copper Handbook Volume VI* the company "holds the Yavapai County rights to the Poehle & Croasland volatilization process, and has a 125-ton reduction plant, blown in, April 1906." It was connected to the P & E by a 2,950 foot long spur. The company claimed the new volatilization process for ore reduction to be satisfactory, but mining customers were skeptical, and no one used the plant. It was taken over by the Mayer Mining and Milling Company without any results. It was eventually dismantled and scrapped. Very few signs of the existence of this plant remain today.

Mineral resources had, and continue to have, a major impact on the

The Rigby Reduction Works were constructed northwest of Mayer in the foothills of the Bradshaw Mountains. A lengthy spur left the P & E just before milepost 25 (from P & E Junction) eight-tenths of a mile from the Mayer Depot. It, too, was unsuccessful and proved to be a disappointment to its investors and the townspeople of Mayer. See also, the map of Mayer in Chapter 9.

A 1941 photograph taken by Rosena Promberger Minucci shows the north end of Mayer with the Red Brick Schoolhouse, the Ladies' Aid Building and a few houses. The site of the Rigby Reduction Works plant is just to the right of the Ladies' Aid Building.

economy of Arizona. According to the Arizona Geologic Survey, Arizona is a leader in United States mineral resources. Every year Arizona leads America in mining production. In 2006, mineral production in Arizona rose to nearly $7,000,000,000. Currently, about 65 percent of the United States' copper is mined in Arizona. Arizona has led the nation in the production of minerals for many years, primarily due to large copper reserves. Significant minable minerals include copper, lead, zinc, gold, silver, manganese, molybdenum, uranium, vanadium and tungsten. Recently, there were more than 70 mining companies doing business in Arizona plus approximately 70 sand and gravel companies. An article by Ryan Randazzo in the Business section of the *Arizona Republic* published on November 14, 2010, states:

> Copper production has been a fixture in the state for 100 years. Arizona's vast copper reserves generate billions of dollars of annual revenue. And with continued investment, the state could remain a major producer of the high-demand metal for years to come.
> In addition to the current mines in Arizona which produce 60 percent of the nation's copper, industry experts said that it was likely that significant undiscovered copper deposits hide underground, and companies should be encouraged to keep up exploration efforts in the state.
> Without continued investment to search for deposits and expand mines, nearly all of the state's copper mines will run out of their existing reserves in about 25 years.
> Most companies digging in Arizona explore enough to replace the copper they dig out each year with new reserves....
> Last year, the copper industry employed 9,100 Arizonans and contributed $9.3 billion to the economy.
> At least three significant new copper mines are proposed in the state, including one—the Resolution Copper Project—that boasts of being the largest copper deposit in the nation. Its output is projected to be more than one-fourth of what the United States produced in 2009.
> With its potentially large reserves, Arizona could remain a lucrative producer of copper for decades to come.

Today, Mayer is no longer a mining town. The closest operating mine in Yavapai County is at Bagdad. But placer mining still goes on, with the gold pan and the sluice box washing and rocking the gravel of the creeks in the Big Bug District. A little lode mining keeps a few mines operating and a few dollars in the pockets of the mine operators. If Joe Mayer was alive in 2011, he would be right out there with the "Big Boys" looking for those new mines and staking his claims.

9

The Railroads Arrive

On September 6, 1898, a special excursion train left the Santa Fe, Prescott and Phoenix Railway Company Depot in Prescott. In the September 7, 1898, issue of the *Arizona Weekly Journal-Miner*, under the headline "President and Mrs. F. M. Murphy Give an Excursion Over This Scenic Line," the reporter wrote:

> President F. M. Murphy of the S F, P & P railroad and of the P & E railroad, with his wife, yesterday, entertained about thirty ladies and gentlemen by giving them an excursion over the new railroad to the end of the track. A special train left the depot at 9 o'clock in the morning, returning at 4 o'clock in the afternoon, the trip being a most delightful one from start to finish.

This shiny, new Brooks Locomotive Company 4-6-0 locomotive (No. 11) was purchased in 1898 for the Prescott & Eastern. It is shown here in the yard at Prescott. Records indicate that it was scrapped in 1927 (Sharlot Hall Museum).

A traveler stands on the Santa Fe, Prescott & Phoenix Railway tracks at P & E Junction about 1900. The spectacular scenery of the Granite Dells and Point of Rocks is in the background.

The new road, although only 26 miles in length, will rival in the beauty of its scenery, the main line, in proportion to its length. From the place where it leaves the SF, P & P six miles from town, the road for about two miles winds its way among the huge mountains of granite rocks, piled up in promiscuous profusion there.

While Wednesday's trip, to many of Mr. Murphy's guests, was their initial one to that section, among the number were several who had made frequent trips by stage along the same route, over that historical and renowned stage line, "The Black Canyon Route." They had made the trip, not as one of pleasure, but as the only means of getting to or away from Prescott, and at times when the passenger who took it might reasonably expect to become the victim of a highway robbery before reaching journey's end. The latter, traveling in a comfortable railway car, over a well constructed and well ballasted railroad, such as this is, the trip was marvelous revelation, and calculated to inspire many reminiscences of former trips when the mention of a railroad through that section would have made one looked upon as a candidate for a madhouse, and the construction of which, even now, is certainly a monument to the energy and enterprise of its promoter and builder.

This was the introduction of the "excursionists" to the then-completed 17 miles of the new Prescott and Eastern Railroad from Prescott to Mayer.

The history of the railroads in Arizona is complicated and fraught with controversy. Battles and political maneuvering between warring railroad companies brought much negative publicity to the prospect of building railroads in the Territory of Arizona

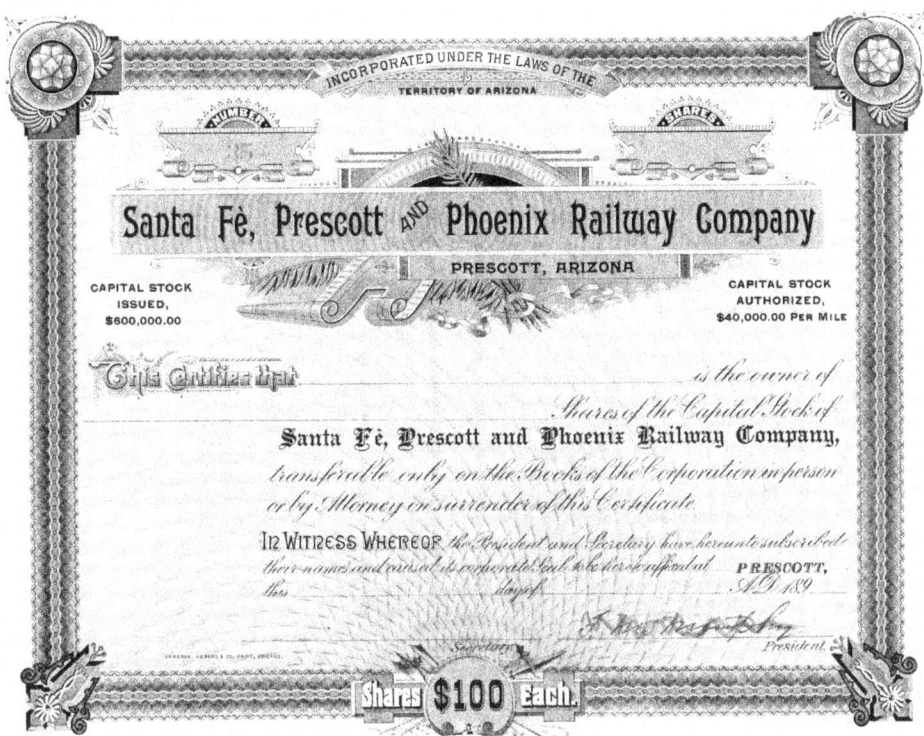

This Santa Fe, Prescott & Phoenix Railway Company of Prescott, Arizona stock certificate number 3 was never issued.

and caused the delay of many proposed railroad projects. However, in spite of the negative connotations of the process, the public and the business owners, particularly those who owned or operated mining businesses, were clamoring for rail service. Various projects started and then ground to a halt. Getting Eastern capital into the Territory in the form of financing for railroads was a goal of all of the various entities interested in building railroads in the West. Financing was a serious issue and getting county bonds approved through the legislature was difficult. Many proposals were made without any results. The people wanted railroads, but they did not want to pay higher taxes to get them. What was to be done?

Entering the picture was Frank M. Murphy, a true 19th century entrepreneur. The older brother of Arizona Governor Nathan Oakes Murphy (the tenth, [1892–93,] and fourteenth, [1898–1902,] Governor of the Territory of Arizona), Frank Murphy came to Arizona Territory in 1878 and was soon considered to be one of the biggest businessmen in Northern Arizona, with interests in the Congress Mine in Congress, the Bashford-Burmister Mercantile Company in Prescott, and, by 1895, the Santa Fe, Prescott and Phoenix Railway (S F, P & P). During Nathan Oakes Murphy's first administration as Governor, Murphy believed that every property owner should pay his fair share of taxes and also recommended that the profits of mines be taxed. On the other hand, Murphy favored exempting new railroads from taxation. His reasoning

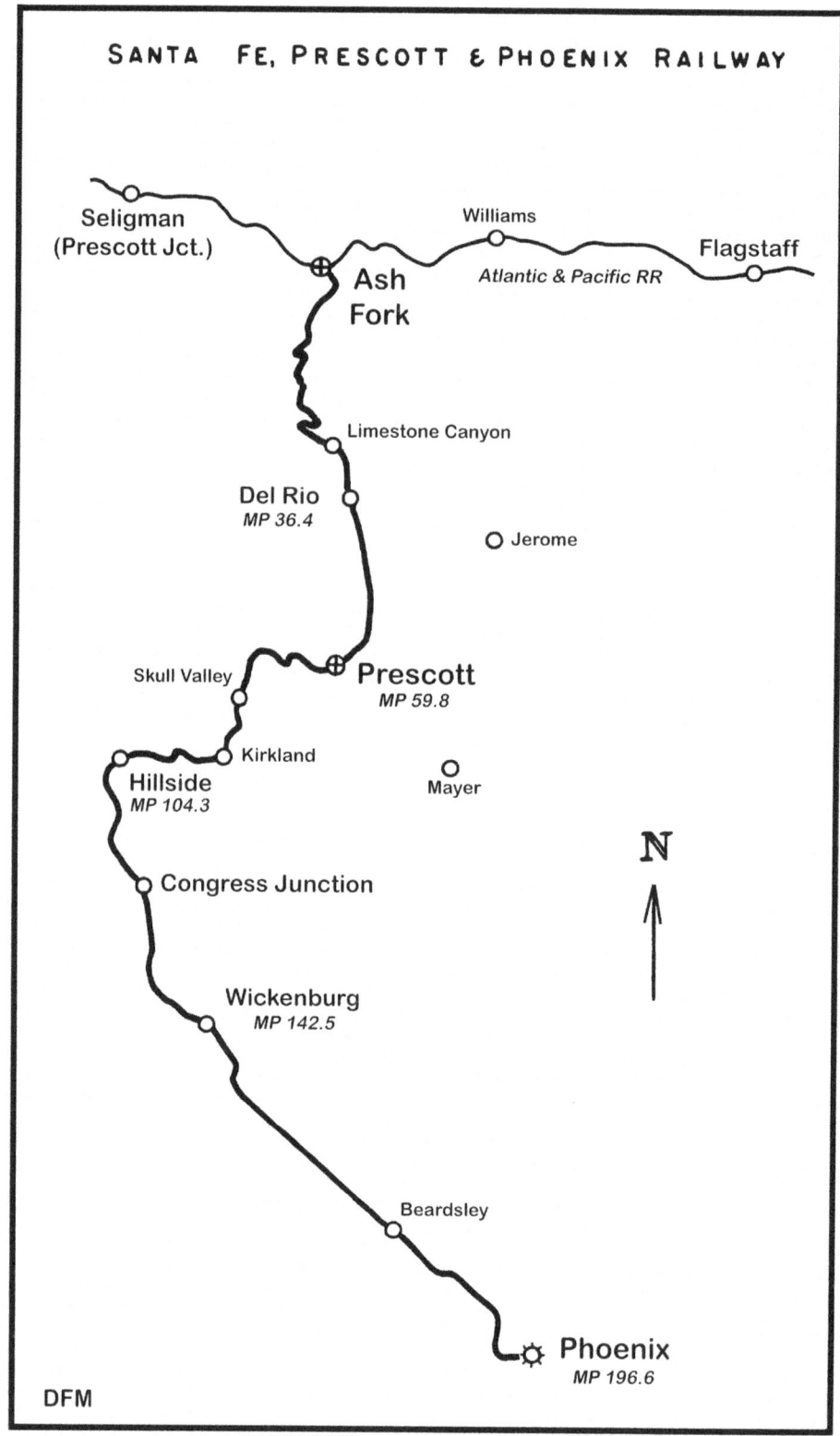

9. The Railroads Arrive

This Brooks Locomotive Company 4-6-0 locomotive and tender (No. 9) was purchased in 1895 for the Santa Fe, Prescott & Phoenix Railway. It is shown here with her crew in Prescott. The headlight is fueled by kerosene. Records indicate that it was scrapped in 1922 (Sharlot Hall Museum).

was that the building of railroads in thinly populated frontiers was a huge financial risk for the builder and a hazardous undertaking. Consequently, one of the laws enacted by the Seventeenth Territorial Assembly did temporarily exempt new railroads from taxation.

In March of 1891, a bill giving 20 years' tax exemption to railroads in the Arizona Territory which began construction within six months had been passed. This tax exemption was not popular with everyone, and consequently, railroads which were constructed under these provisions were often referred to not by name, but rather as the "taxless." An example is an undated newspaper article reporting that "Conductor McRae is laid up at his home from injuries received by falling off a taxless box car at Mayer last Sunday." The term was used in all areas of the Territory and was not reserved for Frank Murphy's railroads. "Taxless" was not meant to be a compliment.

This tax exemption, and others, was the beginning of the railroad boom in Arizona. They were of great advantage to Frank M. Murphy, although it took awhile to get things going. By May 27, 1891, Murphy and his associates had incorporated the Santa Fe, Prescott and Phoenix Railway Company, in which the Atchison, Topeka and Santa Fe Railway Company invested heavily. In April 1893, the S F, P & P Railway, which was commonly known as the "Peavine" due to its winding and meandering route from Ash Fork to Prescott and then on to Phoenix, arrived in Prescott from its connection with the Atlantic & Pacific at Ash Fork. By March of 1895, it had reached Phoenix.

Oppposite: This map shows the route of the Santa Fe, Prescott & Phoenix Railway from Ash Fork to Prescott and on to Skull Valley, Kirkland, Hillside, Congress Junction (now Congress), Wickenburg and Phoenix, a distance of almost 200 miles. Map courtesy of David Myrick, **Santa Fe to Phoenix, Railroads of Arizona Volume 5.**

J. E. Addicks photographed this six-hitch team at Mayer in 1903. To make the mines profitable, a cheaper and more efficient mode of transportation was necessary, and the railroad met that need in most cases. However, equipment still had to get from the railroad to the mines which often required old-fashioned horse power.

Frank M. Murphy, with his S F, P & P Railway line from Ash Fork to Phoenix completed and operating successfully, cast his eyes south toward a line from Prescott to the mining district of Big Bug. A friend of Joe Mayer's, Murphy had mining interests in the Big Bug District, and Joe Mayer was very interested in having Mayer be the shipping point for the entire mining district. The key to developing the mines in the area was a cheaper mode of transportation for shipping ore and receiving supplies, including large and heavy mining equipment. As to Joe Mayer's desire for the railroad to come to Mayer, as reported in the March 19, 1900, issue of the *Arizona Weekly Miner*, "Thousands of admirers whispered to him that the railroad would wipe him out of existence, and that his fate would be told in the vernacular of the Hassayampa 'has beens.' However, Joe Mayer kept pace with the times and ... so has he, in a great measure, been likewise benefitted."

Murphy incorporated the Prescott and Eastern (P & E) Railroad on September 14, 1897. It appeared that little progress was being made on the project, although surveys were completed by S F, P & P Chief Engineer W. A. Drake and his crew in January 1898. But on February 28, 1898, bids were opened at Murphy's Prescott office and the contract for the Prescott and Eastern Railroad was awarded to the Minnesota and Arizona Construction Company of Faribault, Minnesota. Active work on the line

9. *The Railroads Arrive* 173

This Prescott & Eastern Railroad Company stock certificate number 3 is unissued. The initial capital stock authorized was $500,000.

began on March 7, 1898, with headquarters for the project to be located at Point of Rocks, where the new line would join the S F, P & P. A week later, there were 70 men doing rock work at the northern end at what would become P & E Junction, and another 33 men at a camp two miles away who were grading with Fresno scrapers and making good progress. In March, a construction camp consisting of outfit cars and tents was set up just east of P & E Junction. Although there was a shortage of labor due to the gold rush in the Klondike and the Spanish American War, which took 200 Arizona volunteers from Prescott on a special train on May 11, 1898, by the middle of May there were 400 men building the P & E. By July 18 the *Arizona Weekly Journal-Miner* reported that 15 to 18 miles of the grade had been completed. Summer rains slowed the progress, but on July 21, with about half of the grading finished, laying of track began. The July 18 issue of the *Arizona Weekly Journal-Miner* further reported that "the work of laying the track ... will be continued until the graders are caught up with or until the track is completed." Ties were six by eight inches and eight feet long. The track was second-hand 56 pound rail from the Santa Fe Pacific Railroad. Track laying progressed quickly and in a month, ten miles of track had been completed. Also, on July 18, the *Arizona Weekly Journal-Miner* reported that "the track layers and bridge builders are 'right on the heels' of the graders and could complete the line within two weeks if the grade was ready for them. The contractors expect to complete the grade within the time of their contract, by September 15, but it will probably be the first of October, or maybe a little after, before the road will be open for traffic."

The Prescott and Eastern Railroad was a branch line which left the S F, P & P near Point of Rocks in the Granite Dells six miles north of Prescott. The station there, which was known as P & E Junction, was built in 1898. Later, when the location was officially given the formal name of "Entro," the depot was replaced by a small, wooden telephone booth. The route from the junction of the S F, P & P and the P & E ran about two miles through Granite Dells and then entered Lonesome Valley. At mile post 7.7, near where the P & E crossed the old wagon road to Jerome, was a station called "Yaeger," after Yeager Canyon. The railroad then crossed Lynx Creek near the stage stop at John Marr's ranch and, crossing the old stage road several times, followed along the foothills of the Agua Fria Valley until it reached Cherry Creek Station at mile post 14.8 (now known as Dewey). Near the Old King Woolsey Ranch at about mile post 16 (at that time known as Bowers' Ranch), on the Old Black Canyon Road, the P & E changed its course and crossed Big Bug Creek. At the time of the opening of the P & E, the next station was Chaparral and then Huron. The summit, at Poland Junction (4,898 feet), presented a 2.1 percent grade to southbound trains and a 3 percent grade for the northbound trains. The line then passed through Curtiss (Arizona City). The terminus, and the last station on the line, was Mayer's, 26.4 miles from P & E Junction.

On October 15, 1898, the *Arizona Republican*, under a headline of "P & E R. R. Jollification" reported that "the new Prescott & Eastern railroad will be formally opened for traffic on October 16 and this day will be made a day of jollification along the new line." After opening the line on October 15, the company staged an elaborate excursion from Prescott to Mayer, with the Prescott Brass Band on board. Unfortunately, Mr. and

A July 10, 1901, Santa Fe, Prescott & Phoenix Railway Timetable lists the stations served along the route from Ash Fork to Phoenix and from Mayer to Prescott. The Mayer to Prescott run on the P & E covered 31 miles in two hours and ten minutes. Stations along the way included Poland, Huron, Chapparral, Smelter Junction (Humboldt), Cherry Creek, Yaeger and P & E Junction, terminating at Prescott.

An Atchison, Topeka and Santa Fe engine with a tender is approaching the P & E Junction heading toward Prescott in Granite Dells. The photograph is dated 1949 (Sharlot Hall Museum).

Mrs. Frank Murphy could not attend, as Mr. Murphy was ill. The fare was $1.25 from Prescott or $6.00 from Phoenix and did not include meals. A subsequent article about the celebration of the opening of the P & E gave great praise to the new line, stating in part:

> The new line of railroad connecting the rich mineral sections of Chaparral, Big Bug, Crown King, Stoddard, Bumble Bee and other points, known as the Prescott & Eastern Railway, is now completed to Mayer, its present terminus. The P & E R. R. was formally opened by the company with a grand excursion to Mayers last Sunday. At 8:30 o'clock A.M. a train of eight coaches, each filled with joyous, mirth-loving men, women and children, pulled out of the

Grading and laying of track on the P & E was a challenge coming out of Granite Dells, where several deep cuts were required. The rock and earth were loosened with black powder, but the debris from the cut had to be cleared by hand and hauled away in wagons. Most of the finish work was done by hand and by horse power. J. E. Addicks photographed this unidentified cut in 1903.

A Santa Fe boxcar sits on the siding next to the Cherry Creek Depot about 1898 with the town of Dewey in the background. Local residents called their town "Dewey" after Admiral Dewey. The depot was moved to Skull Valley in 1926 and today serves as the Skull Valley Museum (Sharlot Hall Museum).

Although this photograph of the Mayer Depot is dated 1960s, it would have been taken quite a bit earlier as the depot is clearly in use and appears to be in pretty good repair. The "express" sign also indicates an earlier date, probably the 1920s. The depot, 20 feet by 64 feet, was built in 1898. The portion that was moved to Phoenix is the right-hand side shown in this photograph starting just to the left of the bay (Sharlot Hall Museum).

9. *The Railroads Arrive*   177

The small Cherry Creek Depot was built in 1898 for $625.00 and was 12 feet by 48 feet. It served the farming community of Dewey. The Dewey bar can be seen in the background to the left with a horse-drawn buggy parked out front. This photograph was probably taken shortly after the depot was completed (Sharlot Hall Museum).

In 1903 J. E. Addicks took this snapshot of downtown Mayer looking east-southeast with a P & E locomotive, tender and three cars in front of the depot. The 24 foot diameter water tank dominates the view toward the depot. The Mayer Hotel is on the right and the Mayer Station on the left behind all of the trees. It appears to be a quiet day in Mayer as a man walks down the middle of the Black Canyon Road.

Prescott Depot. The first stop was made at the junction, a most picturesque spot among the granite rocks, six miles below Prescott. Here, the company has erected a handsome depot for the accommodation of its patrons, surrounded by the towering peaks of the "granite rocks." Leaving the junction, the road skirts the base of the range for a few miles where it emerges in Lonesome valley down which it follows to the old King Woolsey Ranch.... A station is here which is called Cherry, which is fully equipped for business. The next place, and perhaps ere long the most important station on the line of

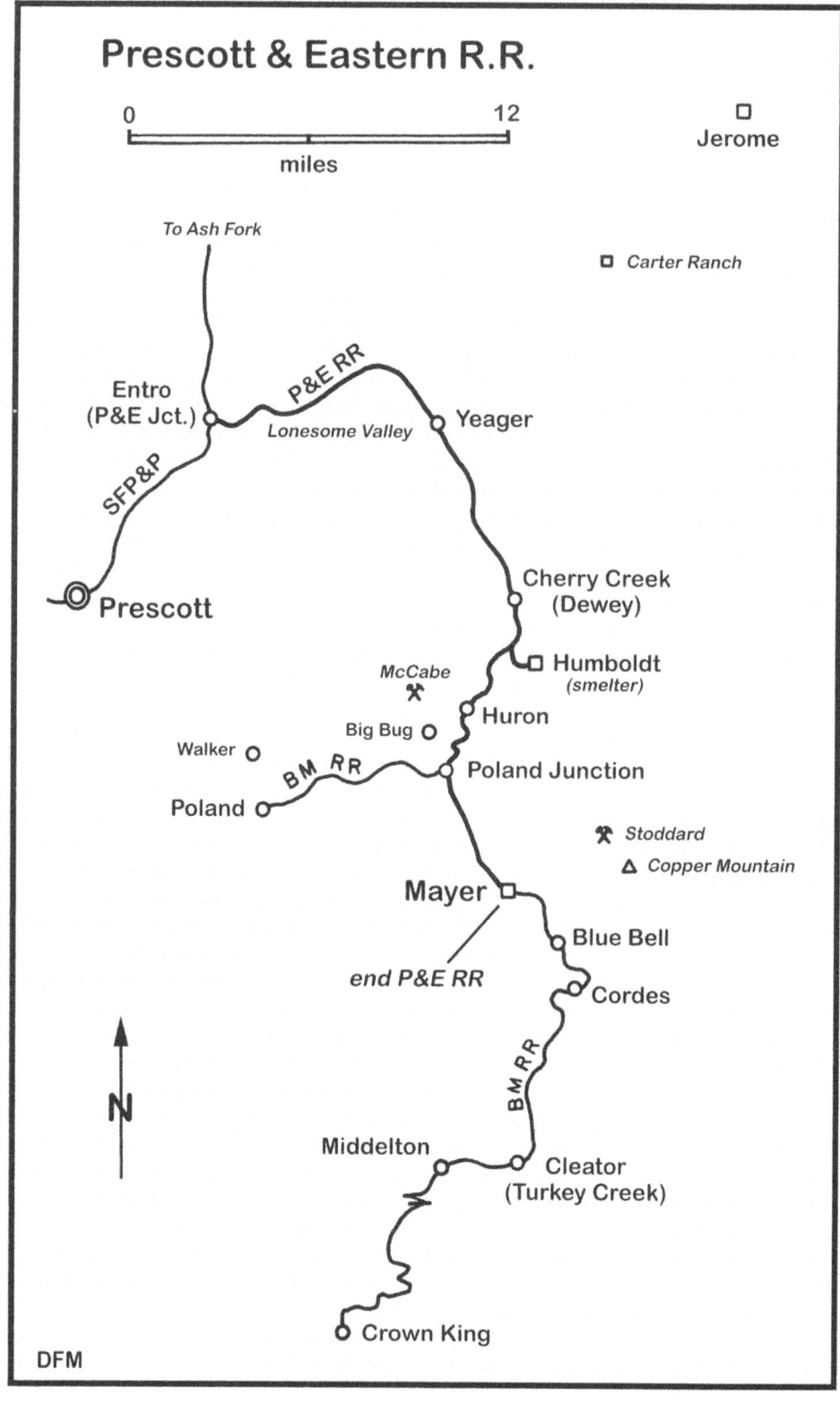

the P & E railroad, is the new town of Huron.... Leaving Huron, the road cuts through a high ridge or spur, which separates Big Bug from the waters and washes emptying into the Agua Fria. From the summit one looks down on Big Bug creek.... About one mile and a half from the summit the road crosses to the south side of Big Bug creek and continues on this side until it reaches Mayer. At Mayer everybody scattered like a flock of quails, to partake of their picnic dinner and to enjoy themselves in several ways.... For three hours the excursionists viewed the famous onyx quarries; but at precisely 1:30 P.M. the locomotive whistle gave a shrill warning, and if one wanted to get home before midnight, that it was time to get aboard the coaches, this was done and our train pulled into Prescott at 3 o'clock P.M.

All expressed themselves as satisfied with their Sunday outing and hoped the company would make these excursions a permanent feature of the road's management.

A short article in the October 19, *Arizona Weekly Journal-Miner* mentioned that "Mr. and Mrs. Joseph Mayer yesterday [October 16] maintained the reputation their place has enjoyed for a long time in the way of setting up the best meals in the Territory. About forty or fifty of the excursionists who failed to take their lunches with them were fed at Mayer's hotel and an elegant meal was served."

The coming of the P & E Railroad sparked the intended growth along the line. The purpose of the P & E Railroad was twofold: first, Frank M. Murphy and others had mining interests in the area, including Murphy's mine at the "new" town of Huron, which was named for the Huron No. 1 and Huron No. 2 mines nearby; and second, everyone, including Murphy and Mayer, wanted to take advantage of the shipping opportunities provided by both the mines in the area and the nearby ranches. The *Arizona Weekly Journal-Miner* reported on October 19 that Jake Miller had made arrangements for a shipment of cattle over the P & E railroad, stating that "he will have the honor of making the first shipment over the road. It will be made next week, just as soon as stockyards for loading can be provided." The Crowned King Mine, by loading a car of ore on the Saturday before the opening excursion, claimed to be the first shipper of ore, which went out on October 17, the Monday after the excursion.

The arrival of the P & E railroad spurred increased production in the area mines and population growth in the camps and towns in the area. According to John W. Sayre, in *Ghost Railroads of Central Arizona*, the area grew in much broader terms than mineral production. Sayre states: "To develop the mines and increase production, more miners were needed. When they arrived, they needed food, housing, and other essentials. Carpenters and merchants arrived to meet the needs of the small camps and towns. Saloons, barber shops, restaurants and boardinghouses soon flourished near the areas of great mining activity. The rapidly growing service sector greatly improved the quality of life in the area. The histories of the communities vary, but the one common factor was the stimulus the railroad provided to their economic and population growth."

About a half mile beyond the Mayer Depot, the Prescott and Eastern Railroad terminated. But, it was not really the end of the line. Built between 1902 and 1904,

Opposite: This map shows the relationship between the Santa Fe, Prescott & Phoenix Railway, the Prescott & Eastern Railroad and the Poland and Crown King branches of the Bradshaw Mountain Railroad. Map courtesy of David Myrick, **Santa Fe to Phoenix, Railroads of Arizona Volume 5.**

The Crown King Branch of the Bradshaw Mountain Railroad had Alco-Brooks 2-8-0 locomotives purchased in 1906. Here, in a snapshot by J. E. Addicks, the locomotive is pushing at least three open cars along the Bradshaw Mountain line.

Another 1903 snapshot by J. E. Addicks along the Bradshaw Mountain line shows the locomotive, again pushing the cars, on one of the many trestles along the line.

the Crown King Branch of the Bradshaw Mountain Railroad started where the P & E stopped and terminated at the mining town of Crown King. One of the most spectacular standard-gauge railroads ever built, it included 28 miles of track which climbed 2,300 feet and included five switchbacks, a tunnel and high trestles. Much of the official grade was 3.5 percent; some was as high as 5 percent, although railroad engineers reported recalling the grades to be as steep as 7 percent. The engineering of this line, known as "Frank Murphy's Impossible Railroad," is credited to John A. Jaeger, a civil engineer with the S F, P & P. Due to the short tail tracks at the end of each switchback, trains were usually limited to an engine, caboose and five freight cars. Unfortunately, by the time this Bradshaw Mountain Railroad was completed, the mines at Crown King were on the decline. The Santa Fe Company sought permission from the Arizona Corporation Commission to terminate service and tear up the tracks in December 1923. The line was abandoned effective March 6, 1926. The Santa Fe, as a parting gesture, ran a special Sunday afternoon excursion train prior to the

This photograph circa 1905 was taken from the Poland Depot looking southeast showing the turntable spur and the turntable. The turntable was 64 feet across and nearly five feet thick. The engineer could pull his train onto either of two spurs, uncouple the cars, turn the locomotive around on the turntable and re-couple the cars for the trip back down the mountain. Engine No. 12 of the Poland Branch of the Bradshaw Mountain Railroad leaves a trail of smoke as she arrives from Poland headed toward the Poland Mill and Poland Junction (Sharlot Hall Museum).

abandonment. The train carried stories, memories and tears as it left the Crown King Depot for Middleton for the last time. Much of the line became an auto road, which is still in use today. Early drivers told of the harrowing trip across the trestles with only inches to spare on each side. Eventually, the trestles were dismantled and the switchbacks and major bridges were circumvented.

Another branch line of the Bradshaw Mountain Railroad left the P & E at Poland Junction northwest of Mayer. The Poland Branch was constructed to reach the Poland Extension Mine from the P & E. In 1901, the 37 mining claims of the Poland mine were acquired by Frank Murphy and his associates, who established the Poland Mining Company. In late April of 1902, the railroad reached Poland, and a month later the first train rolled into town. Since there was no room for a wye, a turntable 64 feet across was constructed next to Big Bug Creek and a bridge was constructed across Big Bug Creek. All trains were led into and out of Poland by a locomotive. A small (12 by 33 feet) depot and platform were constructed on a spur. Private entrepreneurs built a saloon and hotel and a boardinghouse a short distance from the depot. Houses and other business were constructed as the population eventually reached about 800. One of the most challenging and unique features of the Poland Branch of the Bradshaw Mountain Railroad was the Walker-Poland Tunnel. The tunnel, 8,071 feet in length, was completed in 1904 at a cost of $500,000. The tunnel was cut from both sides of the mountain, and when the sections met, they were less than six inches off. The tunnel gave access to the railroad from the Lynx Creek Mining District around Walker.

Georgiana White Johnston and her daughter Mildred pause for a photograph with a hand car at the Poland entrance to the Walker-Poland Tunnel about 1905. The unusual one-person hand car provided transportation through the tunnel. The operator pumped his way from one end of the tunnel to the other (Sharlot Hall Museum).

As with many communities who "hang their hat" on the mining industry, businesses and the population came and went several times. After 1907, mining activity declined significantly. Finally, after a number of starts and stops of mining activity in the area, on July 16, 1919, the line was officially abandoned by the Atchison, Topeka and Santa Fe Railway Company. The Arizona Corporation Commission approved abandonment of the line effective August 1, 1920. The Poland Branch of the Bradshaw Mountain Railway was officially finished. The Santa Fe removed the tracks in 1932. As with the Crown King Branch of the Bradshaw Mountain Railway, the old railroad bed became an automobile road. In the 1940s, the ghost town of Poland was revived and subdivided for summer cabins, today known as "Breezy Pines." It remains one of the most scenic places in Arizona.

One of the businesses which became viable with the arrival of the railroad was the old onyx mine partially previously owned by Joe Mayer. In 1890, William O. "Buckey" O'Neill bought an interest in 40 acres of Mayer's mine, which O'Neill sold three years later. In June of 1897, operating as the Arizona Onyx Company, active production under manager George C. Underhill and his son sparked an onyx mining boom in Mayer. Once the railroad arrived in Mayer, production was increased as transportation of the onyx was eased. Under the ownership of the International Onyx and Marble Company, which bought 100 acres from the Arizona Onyx Company, four carloads of onyx were dispatched to customers as soon as the P & E could ship.

Shortly thereafter, another load was shipped, destination, London. The onyx mining business continued intermittently in Mayer throughout the 20th century and is still in business today.

When the P & E arrived in Mayer, a building boom occurred, mainly by the P & E itself. Looking southeast today from the Mayer Business Block on Central Avenue, the water tank, depot, section house, warehouses and bunkhouse would be to the right (southeast). The water tank was originally constructed southeast of what is now the corner of Central Avenue and Oak Street, near the current location of the United States Post Office. The depot and platform were to the southeast with the main line track and the spur track on each side of the depot. There were three P & E warehouses, a section house, stockyards and, just beyond the switch where the spur re-joined the main line, a section workers' bunkhouse. Once the railroad arrived the stockyards were moved next to the spur track right in downtown Mayer, but, eventually, Mayerites complained about the herds of cattle and sheep and they were moved further away.

Beyond the P & E bunkhouse was a spur to the Treadwell Smelter. The spur was about 700 feet long, and a storage shed and small loading platform were the only structures along the spur owned by the railroad. The Treadwell Smelter was unsuccessful, and, after a new smelter, the Great Western Smelter, was completed in 1915 during a burst of mining activity spurred by World War I, the Great Western went silent and the spur never received much traffic. In 1916, the Big Ledge Copper Company took over the old Treadwell property, but it, too, was unsuccessful.

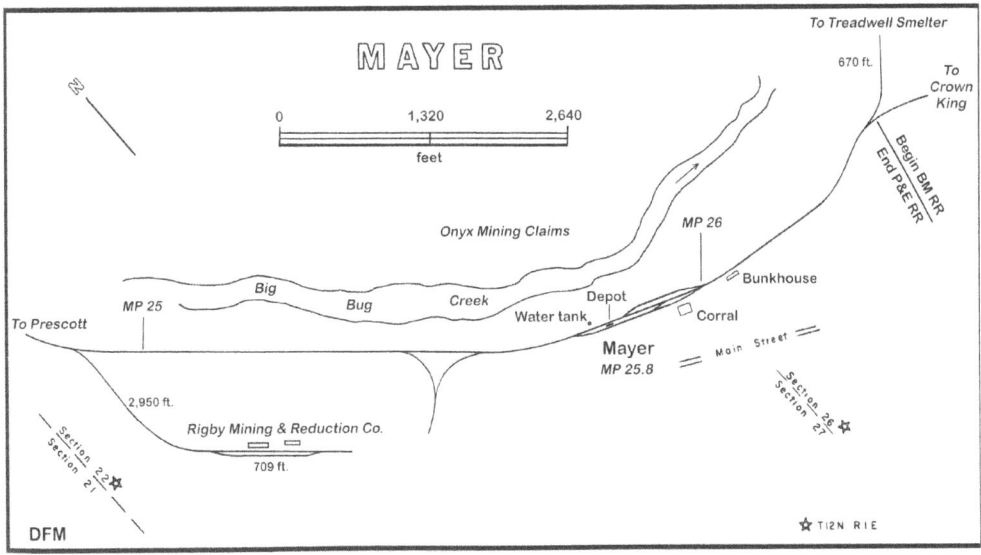

This map clearly shows the layout of the Prescott & Eastern through Mayer, with both the Rigby Reduction Works spur and the Treadwell Smelter spur as well as the terminus of the P & E and the beginning of the Bradshaw Mountain Railroad, the location of the water tank and the depot, the corrals (stockyards) and the P & E bunkhouse. The Black Canyon Road ran between Big Bug Creek and the railroad line. Map courtesy of David Myrick, **Santa Fe to Phoenix, Railroads of Arizona Volume 5.**

The Rigby Mining and Reduction Company works dominated the landscape of upper Mayer on the northwest end of town, but the 125-ton smelter was short-lived. Blown in in April of 1906, it never processed much ore and was soon dismantled. The main line of the P & E runs in front of the smelter in this photograph (Sharlot Hall Museum).

A spur on the north end of Mayer led to the Rigby Mining & Reduction Company. The plant was located on the slope above Big Bug Creek to the north and west of downtown Mayer. The spur, 2,950 feet long, served the Rigby Reduction Works smelter, which was blown in during April 1906. The spur was privately owned and was maintained by the mining company. Unfortunately, the Rigby Reduction Works never processed much ore, and once the big smelter at Val Verde (later called Humboldt) was successful, business was minimal and, eventually, the buildings and machinery were salvaged and removed.

Early on, Mayer was a busy shipping point for ore, livestock and other locally produced goods. Starting in the 1890s, when a road was graded from Mayer to Crown King, Mayer became the shipping point for freight to and from the mines. It was the jumping-off point for the Bradshaw Mountains and its numerous camps and mining activities. Even before the railroad arrived, the small town of Mayer was key to the freighting of goods in the area. Once the railroad arrived, however, its importance as a distribution center, shipping point and passenger embarkation and departure point increased substantially. Less than one year after its opening, a September 1899 article in the *Arizona Weekly Journal-Miner* summed up the successful prospects for the P & E: "Should anyone doubt the advisability of building the Prescott & Eastern railroad, a trip from Prescott here will convince them of the wisdom of the enterprise. Passenger

9. The Railroads Arrive

This photograph shows a group of Yavapai Indians on the platform of the Mayer Depot, probably about 1905. The men are all seated on the bench. The women, who are standing with the children, are wearing their traditional dress of the time. The sender, Carl, wrote, "Dear Mother—here is a picture of the Mayer Depot and some of my lady friends." He mailed the postcard October 2, 1906, from Mayer. Real photograph postcard.

Something big must be going on in Mayer in this photograph dated January 7, 1907. Most of the population of Mayer must be at the Mayer Depot. The horse and wagon are decorated with American flags. At least three cars make up the consist.

traffic is better than one would naturally suppose, while the freight business is enormous. Some of the old Prescott fossils should make the trip for enlightenment."

Five years of rapid growth followed the arrival of the P & E. From all reports, the accommodations at Mayer, which of, course, included the Mayer Hotel and Joe Mayer's mercantile store, were comparable to those in other parts of the Territory. A lengthy article in the September 11, 1902, issue of the *Daily Journal-Miner* reported, in part, on the post-railroad conditions at Mayer and the hospitable reputation of Joe Mayer and his town, stating:

> Mayer, located further down Big Bug, is likewise favorably situated as a center, and seems to be enjoying a favorable degree of prosperity. This latter place enjoys distinction and renown as a hospitable center, and cherishes the same regard today as it did in the early eighties when it was "the only" stage station in Arizona that seemed to regard the wayfarer with any degree of courtesy or comfort, aside from a pecuniary consideration in return. The old time characteristics of Joe Mayer seem not to be shattered by the iron horse coming along, and whether a man is behind a burro or a locomotive, he plays the same system today that he did twenty years ago, giving the best that money can buy and never gets red in the face when held up or stood off. He has become so famous as a host that he can have anything he wants in Yavapai.... He is building up what promises to be a strong commercial center, and there is but little doubt that with the contemplated improvements, the town that bears his name will become prominently identified to this section of Arizona as it is in the center of a mining region of unquestioned richness.

Prior to the arrival of the P & E in 1898, Huron had just one house and one tent. One month after the railroad came to "town," there were ten buildings. The saloon is conveniently located just a few steps away from the railroad tracks. This photograph of Huron is dated 1902 (Sharlot Hall Museum).

Frank Murphy, along with E. B. Gage of Prescott, had secured rights to the Huron Mine, which was then only a stone's throw from the railroad. Encouraged by the supposed gold and copper potential of the Huron Mine, which had been staked by a "tenderfoot" in 1896, two years before the arrival of the P & E, Murphy and Gage hoped to reap considerable minerals from the mine. However, it never really showed the "color" that they hoped. In the process, however, a community was established by the railroad with a small depot, warehouse and siding near the mine. The Huron Mine, which prior to the arrival of the railroad sported one building and one tent, had ten buildings one month after the arrival of the P & E. Eventually, a post office was established in August of 1901. In *Ghost Railroads of Central Arizona*, John W. Sayre includes an excellent history of the town of Huron, stating, in part:

> In addition to the saloon and the concerns of Wingfield, other businesses were represented in Huron. The P & E Ry. was not the only large company with an office in the community. When the railway began operations in Huron, the Western Union Telegraph Company and the Wells Fargo opened offices in the depot. The small railroad depot, which only measured sixty-four by twenty feet, was not cramped for space as the same agent served all three companies.
>
> Essential public utilities and services were available in Huron. A small post office, established in 1901, served the community for more than a quarter of a century. Telephone service was provided by the Prescott Electric Company and connected Huron with most camps in the area as well as the social and cultural center of Prescott.... The county created the Huron School District in 1902. It remained active until 1904 when, due to small enrollment, it was absorbed into the McCabe School District.
>
> Weeds, rotted wood and rusted rail soon marked the site of the little town. In the late twenties, the railroad dismantled its depot and warehouse for tax purposes. It removed its rail and legally abandoned its line through Huron in 1958. The eroded railroad grade that slowly winds its way uphill toward the silent smelter town of Humboldt and downhill toward the small settlement started by Joe Mayer holds the heart of a bygone day. The town of Huron, save the old depot grade and brush covered barrel hoops, has been claimed by the Big Bug countryside.

But the town of Mayer had a different outcome. In an article in the March 1900 issue of the *Arizona Weekly Journal-Miner*, under the headline of "An Historical Locality That is Holding its Own as an Active Industrial Center," with the sub-headline of "Like In Pioneer Days it Remains in the Saddle and Has a Bright Future Ahead," the enthusiastic reporter wrote in part:

> Probably in all Arizona there is no section more familiar or agreeably known to the masses than can be said of Mayer. In the earlier days of staging from Prescott to Phoenix it was regarded as a luxury when the wild mustangs pulled up at the place, and after leaving it the universal verdict was that in hospitality and accommodations, it savored of the delightful and refreshing indeed. So famous had this center become in the annals of overland travel, that the monotony of riding on the hurricane deck of a prairie schooner was more or less of a pleasant anticipation after one had partaken there of the rough and ready generosity, to await him at every turn. With the decline of the stagecoach and the building of the railroad, however, one would naturally believe that the old landmark of the "fast and furious" was to be supplanted by a delicacy of regard that the tenderfoot era was sure to unravel, or, in other words, the place was to lose its old-time identity and charm. But such is not the case.... Mayer of old is still the same, but as the boys say, on a bigger scale.

The Mayer Owl and Lev Nellis' Cash Meat Market are shown from the east side of the P & E track in the 1920s. All of the cars in Mayer must be at the Owl. Perhaps it is movie day. A bullet-riddled switch stand and target stands as a sentry on the left.

On December 31, 1911, the corporate entity of the Prescott and Eastern Railroad ended when it was sold to the California, Arizona and Santa Fe Railway Company. On August 1, 1920, the Atchison, Topeka and Santa Fe Railway completed a merger of the Santa Fe, Prescott and Phoenix, the Prescott and Eastern and the Bradshaw Mountain railways, which they had purchased in 1901. This was, unfortunately, the beginning of the end for these small, regional railroads, and within the time span of the next several decades, they were gone.

A boom in the railroad business on the P & E occurred in 1956 when an oil pipeline contractor shipped 219 cars of fittings and pipe to Mayer, but this was a one-time bonanza and predictions for future traffic on the line looked bleak. In 1958, the railroad abandoned its right-of-way through Mayer and removed the tracks, including

A boxcar sits on the siding at the back of the Mayer Depot in a photograph taken by Rosena Promberger Minucci in 1941. Seventeen years later, the tracks were removed and the P & E was no more.

The Mayer Depot as it was being prepared for its move to Phoenix in 1962. Vandals and the weather have taken their toll (Sharlot Hall Museum).

the Great Western spur. As had been agreed at the time the right-of-way was acquired, the abandoned right-of-way was returned to its previous owners. Subsequently, some of the right-of-way was acquired by the Mayer School District and others. The Mayer community center was built on a portion of the abandoned right-of-way. However, the fate of the Mayer Depot appeared to be demolition by neglect. The baggage room and platform were torn down and the depot was left to the ravages of vandals and the elements; that is, until Arizona historian, newspaper columnist and author Don Dedera discovered the depot. An article in the March 12, 1959, issue of *The Yavapai County Messenger* included a photograph of the Mayer Depot. The caption read: "No longer is the cry 'all aboard' heard at the Mayer railroad station.... Long since boarded up and tracks removed, the building is still a memento of the past and also serves to shelter burros and cattle from the hot sun." Dedera decided to purchase the Mayer Depot, move it to Phoenix and incorporate it into a new house he wanted to build. While someone else had purchased the depot and had moved it to another location in Mayer, Dedera was determined, and eventually made arrangements to buy the depot. On October 15, 1962, the deal was done.

At about 3:00 P.M. on Monday, March 12, 1963, the Mayer Depot left Mayer on its way to Phoenix. The cavalcade was led by Dedera as the depot started south under the power of the Arizona House Moving Company. *Prescott Evening Courier* correspondent Jack Roth reported that "after 64 years, the red painted passenger station started on its journey to the south with her head held high and proud. But, it may not be lost to Mayer forever, for Dedera owns property in Mayer and there is just a chance he may someday bring it back to its birthday town and set it down, not too far from the spot where it has served so well." But that was not to be. The depot consisted of three rooms: the Station Master's office; the public lobby, or waiting room; and the baggage handling room. The building totaled about 750 square feet. The Mayer Depot made its journey to Phoenix down the Old Black Canyon Highway. At over forty feet in height on the truck, the little depot encountered many telephone and electrical lines. Once it reached Phoenix, to save some time and avoid further lines, Dedera made arrangements to take the depot down the middle of the runway at Deer Valley Airport. The depot was then stored for a time while Dedera worked on it, restoring the windows, repairing damage, replacing the pine siding with redwood and putting it right. And then, in early 1964, the Mayer Depot went traveling again, making its way up the street to its new home, and, with the assistance of two huge cranes, the Mayer Depot was set down on top of a hill in northeast Phoenix. One of the Dedera's neighbors, who was sitting on his front porch as the depot went by, told him that "in all of his 60 years as a railroad engineer, he had passed by many a depot, but he had never had one pass by him." The Mayer Depot was to become a part of a new home designed by Phoenix architect Fred Guirey and built by the Dederas. The historic integrity of the depot was retained, including the waiting room, agent's office and ticket window. Freight storage areas were incorporated into a bathroom, den, workshop and darkroom. The Mayer Depot became a beloved part of railroad aficionado Dedera's home. Although Don Dedera sold the home in 1969, a year later Edith and Don Kunz purchased the home and raised their family there. Over the years and several owners, changes to the exterior have been minimal. Today, the Mayer Depot House has recently

The Mayer Depot as part of Don Dedera's home in Phoenix. Dedera did a monumental amount of work to restore the depot and incorporate it into his home. The rest of the house was designed to complement the architectural style of the Mayer Depot. Photograph by Nancy Burgess, 2009.

The current owners of the "Depot House" have restored the depot portion of the building with original details, including the ticket window, shown here. Photograph by Nancy Burgess, 2009.

The Depot House, shown from the Mayer Depot (north) end of the building, on its hilltop in northeast Phoenix. It has recently been remodeled and restored by owners Will Auther and Allison Wiener. Photograph by Nancy Burgess, 2009.

been restored and upgraded under the guidance of architect Joe Cook, who once worked for Fred Guirey, contractor Eric Workman and interior designer Terese Messina by the current owners, William F. Auther and Allison Wiener. The Mayer Depot portion of the house was further restored during the project. Mr. Auther has turned the Station Agent's office into his Mayer Depot memorabilia room and has developed a comprehensive timeline and history of the Mayer Depot. The house was a feature of the Sunnyslope Historic Home Tour in 2010. The Mayer Depot lives on in its current role as part of a much beloved home on a hilltop about 75 miles from its original site.

10

The Depression Years and World War II in Mayer

One of the most significant indications of progress in the Arizona Territory in the 20th century was that Arizona finally obtained statehood on February 14, 1912. A 30-plus year effort to become a state finally paid off when President William Howard Taft signed the statehood bill on that day, ushering in a new era for Arizona. In 1891, the Territory had elected 21 delegates to a Constitutional Convention. Outraged over the admission of states with less population than Arizona Territory, and having worked hard since the first statehood bill for Arizona Territory had been introduced in 1883, the delegates drafted a constitution, and the voters of the Territory of Arizona ratified it. There was great excitement and expectation that the United States Congress would approve it. Governor Nathan Oakes Murphy personally went to Washington, D. C. in 1893 to make the plea of the citizens of Arizona Territory for statehood. But his pleas were ignored and nothing happened. Subsequent appeals also fell on deaf ears.

A crowd gathered at the Prescott Santa Fe Depot for the visit of President William Howard Taft in 1918. Taft also visited Arizona in 1909, prior to his approval of the Statehood Enabling Act for Arizona in 1910 (Sharlot Hall Museum).

Finally, in 1906, an election was held in which the voters of New Mexico and Arizona Territories could let their wishes be known as to their desire for combined or separate statehood. Arizona voted 3,141 in favor of joint statehood and 16,265 against. New Mexico voters disagreed, voting 26,296 in favor of joint statehood and 14,736 against. In the overall count, those in favor of separate states won by a margin of 1,564 votes. Once the outcome of the election was announced on November 7, the citizens of Douglas, Arizona, held a formal funeral for "joint statehood." But Washington, D. C.'s politicians were unmoved. It wasn't until President Taft visited Arizona Territory three years later in 1909 that the ball started rolling. The House of Representatives finally passed the statehood enabling act for Arizona and New Mexico as separate states in January 1910 and sent it on to the Senate. Arizona celebrated, but there was a long road ahead. President Taft did not like the proposed constitutional provisions allowing for the recall of elected officials, especially judges, so he rejected it. After a vote of approval in the Senate on June 16, Taft signed the statehood enabling act on June 20, 1910, finally putting Arizona Territory on the road to statehood.

Delegates to another Constitutional Convention were elected in September 1910. Surprisingly, the Democrats came out in the majority, and Governor Richard Sloan predicted that statehood for Arizona, so recently a sure thing, was now a remote possibility, declaring that Arizona's chances for statehood were "about the same as for annexation to the Russian Empire." A constitution was produced in December 1910, but there was still the "problem" of the recall provisions in Arizona Territory's "wildly radical" constitution. Arizona voters wanted recall, but they wanted statehood more, so after a compromise regarding the recall provisions, the voters of Arizona went back to the polls and approved the constitution without recall. In November 1911, the *Florence Blade-Tribune* (Florence, Arizona's newspaper) published the following poem by Colonel Thomas Weedin*:

> We will tolerate your gall
> And surrender our recall
> Till safe within the statehood stall
> Billy Taft, Billy Taft
>
> Then we'll fairly drive you daft
> With the ring of our horse laugh
> Billy Taft, Billy Taft
>
> As we joyously re-install
> By the vote of one and all
> That ever glorious recall
> Billy Taft, Billy Taft

Arizona voters knew that they could reinstate the controversial recall as soon as statehood was granted. Taft knew this also, but felt he had a duty to see that statehood was not granted until the Arizona constitution was void of recall provisions for elected officials. Finally, on Wednesday, February 14, 1912, President Taft affixed his signature to the Arizona statehood proclamation. Valentine's Day 1912 was considered by the voters of the new State of Arizona to be the day they were "emancipated from nearly

---

*Weedin ran against George W. P. Hunt in the 1911 Arizona Territorial Gubernatorial Primary election and lost.

10. The Depression Years and World War II in Mayer

This overview of Prescott dates from about 1907. The message on the reverse of the postcard reads, "Our mountain city of Prescott, Ariz. taken from the reservoir on the south." The Courthouse Plaza is right of center and Granite Mountain provides the background. The postcard was mailed in September 1907. Real photograph postcard.

half a century of territorial bondage." The February 14, 1912, issue of the *Arizona Republican* proclaimed "The Forty-eighth State Steps Into the Union Today." That day, Governor George W. P. Hunt was sworn in on the balcony of the Capitol Building in Phoenix. As Don Dedera wrote in *Arizona Highways Album, The Road to Statehood:* "Awaiting the flash from Washington were a people made snappishly independent by federal abuse and neglect. They had not only endured Geronimo, confiscatory freight rates, and libelous journalism, but a Congress that had seriously proposed naming the Territory 'Gadsonia' after the Gadsden Purchase. Rallying around the poetry of Sharlot Hall, they had noisily defeated joint statehood with New Mexico. And they had gambled the whole game by adopting the nation's most progressive constitution. This was the moment of a thousand victories." Finally, the celebrations could begin. In Prescott, the Yavapai County seat, a throng gathered at the Courthouse Plaza to witness the planting of a white oak "Statehood Tree," listen to church bells ring and hear orations. As soon as statehood was granted, Arizona revised its constitution to include the provision which was so disliked by President Taft allowing for the recall of elected officials *and* giving the vote to women.

Meanwhile, the boom years of the first decade of the 20th century in Yavapai County waxed and waned from about 1910 through the mid-teens. Smelters were constructed in Mayer and in Humboldt, and new methods of separating the precious metals from the earth were constantly being developed and tried, but often without success. An article in the September 11, 1902, issue of the *Daily Journal-Miner*, sub-headlined

Another day for a celebration in Prescott was the unveiling of the Roughrider Memorial Statue on the Courthouse Plaza on July 4, 1907. The sculpture, by Solon Borglum, is considered to be one of the finest equestrian statues in the United States. A large crowd is anxiously awaiting the removal of the covering. A small sign which reads "keep off the grass" is having no effect.

Great excitement and celebration broke out all over the Territory of Arizona on February 14, 1912. Arizona had finally gained statehood. In this photograph, the citizens of Prescott and their neighbors have gathered on Admission Day for the planting of the Statehood Tree. Unfortunately, this tree did not survive. However a Deodora Cedar which was planted on July 4, 1910, in recognition of the fact that the U. S. Congress had agreed that Arizona and New Mexico Territories should be recommended for statehood became the "Statehood Tree." It still graces the Courthouse Plaza (Sharlot Hall Museum).

Livestock contributed to the economic base of the Santa Fe, Prescott & Phoenix Railway and the Prescott & Eastern Railroad. Here, cattle are being herded into the loading chute for shipment in the stock cars in the background. This photograph is circa 1910 (Sharlot Hall Museum).

Red Cross Day in Mayer was held on May 24, 1918. The celebration included a midway, floats, carnival, parade, barbeque, the Prescott City band and a game called "Kill the Kaiser." An article in the **Prescott Journal-Miner** *on May 25 mentioned that "the high school cadets were a wonder not only to those from outside Prescott but also to the hundreds of Prescott people who were there." This photograph may show the cadets lined up in front of the Mayer Business Block. Proceeds from the event were donated to the Red Cross. Real photograph postcard.*

"Big Bug and Its Bright Outlook," praised the advances made in Mayer since the arrival of the Prescott and Eastern Railroad in 1898, stating: "Mayer, located further down the Big Bug, is likewise favorably situated as a center, and seems to be enjoying a comparative degree of prosperity. The latter place enjoys distinction and renown as a hospitable center, and cherishes the same regard today as it did in the early eighties."

But this prosperity was not to continue. Mining enterprises were plagued by smelter failures, fires, and labor issues. On July 4, 1907, Yavapai County residents gathered at the Courthouse Plaza in Prescott to celebrate the unveiling of the "Rough Rider Memorial Statue" by sculptor Solon Borglum. By 1910, most of the profitable mines in the area were played out or were curtailing operations. Some were shutting down. The local economy was down again. However, by 1912, things looked somewhat brighter in the new State of Arizona. Railroading was still big business; the mines were pouring forth their copper; lumbering was thriving; cattle ranching was booming; tourism was gaining ground and new tourist resorts were being built; and business was making healthy gains. Arizonans were mad for the automobile, which cut their travel times from days to hours where there were roads. In 1914, Arizona "went dry," passing prohibition, which essentially made liquor just a little harder and less convenient to get. But there was other trouble on the horizon.

The start of World War I in 1914 eventually brought great changes to the nation and to Arizona. Once the United States entered the war in April 1917, Arizona became a vast training ground for the war. World War I gave a huge boost to the economy of Arizona, particularly mining, partially due to subsidies paid by the United States government in the form of price supports—which, in turn, gave a boost to any and all

This view of downtown Mayer shows the Mayer Hotel on the right, the Mayer Business Block and the Owl on the left and the Mayer Depot in the center background. There are a number of cars on the street, plus horse-drawn wagons. This image would have been taken in the 1920s. Real photograph postcard.

businesses associated with mining: railroads, mercantile, and manufacturing. Copper prices increased substantially, and where this metal had often been ignored for gold and silver, it was now a viable commodity worth recovering. Ores which were previously deemed to be too low-grade to be profitable soon became very profitable. Copper, gold and silver are often found in the same veins, and tailings from which the gold and silver had been removed could yield significant amounts of copper. In turn, copper veins often yielded viable amounts of gold and silver. All of these minerals, especially copper, were in great demand during World War I, and production was greatly increased. The years of 1917 and 1918 were particularly bountiful. However, a miners' strike in 1917 put a damper on the mining industry statewide and led to violence. Local citizens, fearing the violence, armed themselves. When the radical Workers of the World seized control of the strike efforts, they soon found themselves being "run out of town on a rail" in Jerome and Bisbee. Arizona's vigilante heritage was not dead.

Due to the mining boom, shipments rose dramatically on the Prescott and Eastern and the Bradshaw Mountain railroads, and the communities these lines served also became much more profitable. However, when World War I was over, the mining boom ended abruptly, and activity at the mines dropped off drastically. Cotton, cattle and copper prices plunged. Revenues for the railroads, mercantile stores and smelters suffered accordingly. Shipments of cattle and sheep kept the railroads in business, but barely. Those railroad lines which relied almost completely on mining activity were impacted the most severely.

The post–World War I mid-1920s saw the dark shadows of the beginning of the Great Depression. A number of banks in Yavapai County closed their doors, and in some cases, the customers lost their deposits. As was mentioned earlier, by 1926 the Mayer State Bank was closed. As Margaret F. Maxwell wrote in her article in the *Journal of Arizona History* entitled "The Depression in Yavapai County," "The effects of the Great Depression were felt later, and perhaps with less severity, in Arizona than in other parts of the United States. The people of Yavapai County in northern Arizona and in Prescott, the county seat, prided themselves on their frontier traditions of thrift, neighborly cooperation and independence from governmental interference." This same philosophy would have been the norm in the more rural areas of the county, such as Mayer, where everyone knew almost everyone, and, consequently, knew almost everyone's personal situation.

The stock market crash of "Black Friday," November 1, 1929, didn't get much press in Arizona, although the *Prescott Courier* ran a small item: "Stock market closed today."

Interestingly, although the population of Arizona grew by nearly one-third between 1920 and 1930, the rural population grew at a larger percentage than the urban population. The total population of Arizona in 1930 of 435,573 persons, up from 334,162 in 1920, placed Arizona as the 43rd of 48 states in population in 1930. However, Arizona was the fifth largest state in area, making it one of the least densely populated. Although the farm population was declining in the 1920s nationwide, in Arizona it was increasing, due mainly to the ability to irrigate land in the Salt River Valley as a consequence of the construction of reclamation dams on the Gila and Salt rivers. Agriculture-based communities in general grew through the 1920s, while the

An overview of Mayer taken from the south end of town looking mostly west. The Red brick Schoolhouse and the Ladies' Aid Building are in the approximate center of the photograph above the large tree in the foreground. The Catholic church is on the left. This photograph was taken in 1934 (Sharlot Hall Museum).

population of other communities for the most part remained steady with slight growth or slight declines. A few key industries dominated the employment in the state. In 1930, agriculture provided the most jobs in Arizona, with manufacturing and mechanical industries coming in second, minerals extraction third and transportation fourth. According to William S. Collins, in *The New Deal in Arizona*, "Between 1929 and 1933, the economy spiraled downward as declines in one industry reverberated in others. The interconnectedness of the economy, industry to industry and region to region, meant that the Great Depression, with few exceptions, was general. Few communities escaped significant unemployment, whether they were unemployed copper miners, construction workers, railroad employees, or those dependent on their incomes such as local merchants." Unfortunately, this scenario fit Mayer and the surrounding camps and small communities exactly and the decline in employment by the railroads was a key factor. Railroads were already experiencing competition from cars and trucks when the Depression struck. The reduced traffic and resultant bankruptcies across the country put even more strain on those communities which relied on the railroads. As Margaret F. Maxwell wrote in "The Depression in Yavapai County," "By mind–September, 1930, a Federal Department of Labor survey of economic conditions in Arizona announced that 'the supply of all classes of labor somewhat exceeded requirements.'" As the Yavapai County Chamber of Commerce, under the direction of Chamber of Commerce Secretary Grace Sparkes, began relief efforts in the county, industrial and political leaders failed to take action, apparently believing that those who were jobless were somehow responsible for their own situations. In the meantime, copper prices dropped so low that the United Verde Copper Company in Jerome announced that profitable operations were no longer possible and announced preparations to close. The effects were felt almost immediately all over Yavapai County. The Great Depression was at hand.

Although a number of federal relief programs came and went, such as FERA, CWA, the NRA, the PWA, the AAA, and, under the "New Deal," the WPA and the NYA, Grace Sparkes was able to implement programs and projects around Yavapai

10. The Depression Years and World War II in Mayer   201

The United Verde Extension Company's mine in Jerome shut down operations in the 1930s when copper prices dropped.

County which put local men to work. But small towns which were unincorporated and had no government infrastructure, such as Mayer, were on the short end of these efforts. However, a program instituted in 1933 by President Franklin Delano Roosevelt did have an impact on Mayer. A bill was passed in the United States Senate giving the president broad powers to employ men in conservation work on both public and private land. The result was the formation of the Civilian Conservation Corps (CCC) in 1933, and the first enrollee signed up only 37 days after Roosevelt's inauguration. As William

The construction of the Yavapai County fairgrounds in Prescott in 1934 put local men to work during the Great Depression (Sharlot Hall Museum).

S. Collins wrote in *The New Deal in Arizona*, "Enrollees were employed and paid by the federal government, their work projects coordinated by the Departments of the Interior and Agriculture, and the Army regulated their living places and off work activities." The quota for Arizona was 1,000 unemployed, single young men (between 17 and 24) to be employed primarily "in the clean atmosphere" of the state's forests. Initially, enrollees signed up for a six-month enlistment and could "re-up" for six more months. After that, they were expected to leave the corps and enter the job market. Workers were to be paid $30 a month by the federal government, $25 of which was to be sent back to their families. Also, discrimination due to race, color or creed was barred by the act creating the CCC, although women were excluded from the program. The CCC quota in Arizona filled quickly, and soon there was a waiting list. A recruit from southern Arizona on his way by train to Fort Huachuca for training wrote home that he and his comrades were a "Forest Army ready to go to war ... We are chugging away towards the front where we will fight hard to win ... best regards from the Fighting Forester."

The fact that much of Arizona consisted of federally owned land made it very desirable for CCC projects and camps, which would be built almost exclusively on federal land. In Arizona, most of the camps would be supervised by the U. S. Forest Service, which reflected the importance of the Forest Service as one of the largest landholders in the state. The second major agency to employ CCC workers was the National Park Service. Their efforts were concentrated at Grand Canyon National Park. In Yavapai County, the CCC, also nicknamed the "Colossal College of Calluses," was involved in building roads, bridges, campgrounds, bath houses, cabins, latrines, fences, cattle guards and sidewalks. For the Forest Service, they built storage facilities for equipment, fought twig blight, built campgrounds and trails, installed stone retaining walls, and refurbished existing buildings. Altogether, 50 camps were constructed in Arizona between 1933 and 1942, when the program ended. One of those camps was constructed at Mayer, Arizona.

The first truck parked in front of the Mayer CCC camp is a U. S. Forest Service vehicle. The camp opened in the fall of 1933. The flagpole is made from a de-barked tree, probably a Ponderosa Pine from the higher elevations of the Bradshaw Mountains.

10. *The Depression Years and World War II in Mayer* 203

The CCC camps in Yavapai County were assigned to the 8th Corps Area out of Phoenix. The Mayer Camp was Camp F-33A for Company 822. Initially, the Mayer camp was a winter camp and the summer camp (F-79) was at Walnut Creek. According to the *Official Annual 1936 of the Phoenix District, 8th Corps Area*, Company 822 was organized on May 29, 1933. In the fall of 1933, the Company moved to winter quarters at "Turkey Creek, Mayer, Arizona." The camp at Mayer was considered to be a permanent camp with "rigid" quarters constructed of wood, as opposed to a "fly" or "spike" camp, which was considered to be temporary and moveable and consisted primarily of tents. The camp was constructed on what had been farm land. The camp was described in a July 1941 inventory as having 26 buildings, including four barracks for

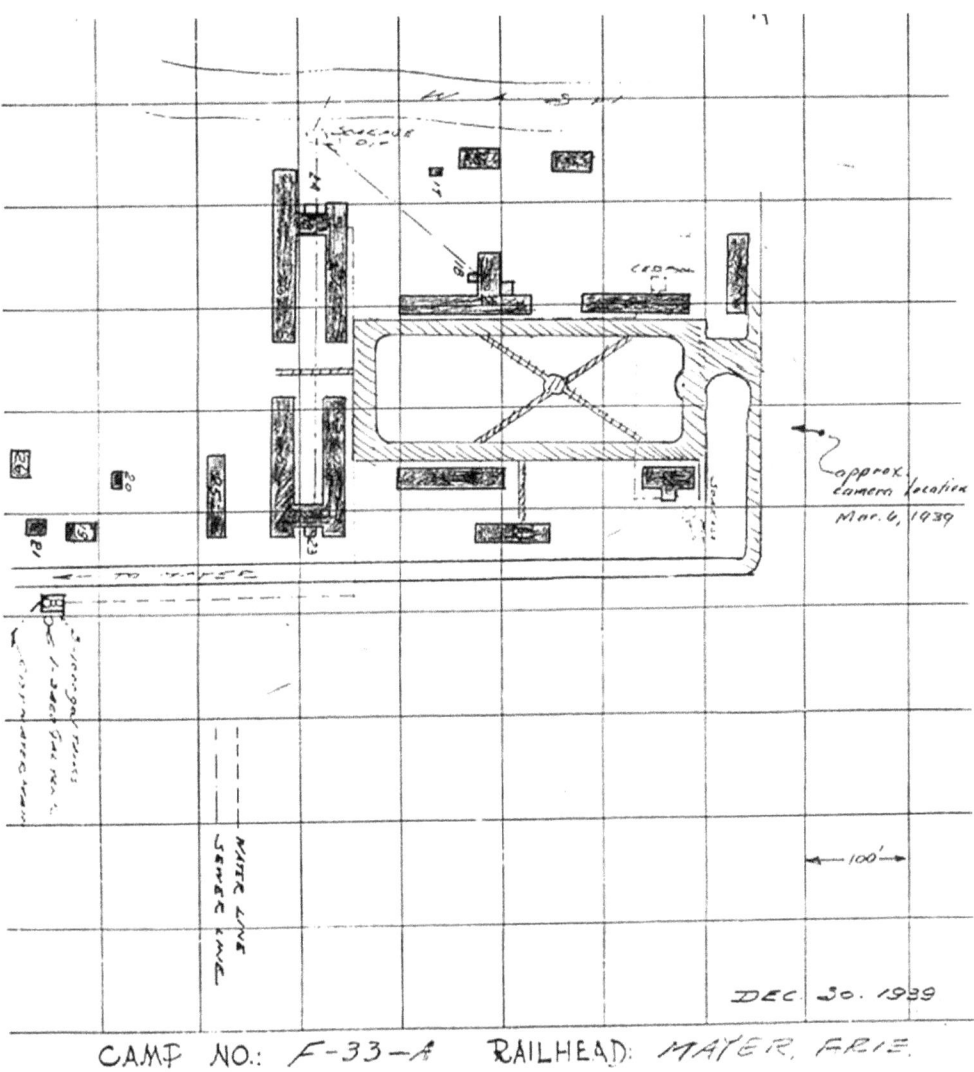

A plan of the Mayer CCC camp F-33-A dated December 30, 1939, shows the buildings (dark color), the parade ground, water and sewer lines and the road to Mayer. The camp complex was about 800 feet by 500 feet.

Recreation was an important part of the CCC experience. Here, a group of CCC enrollees enjoy a picnic in the pines near Prescott on April 30, 1937. Photograph by Bate (Sharlot Hall Museum).

A panorama of the Mayer CCC camp F-33-A taken circa 1936 with the camera facing west shows the entrance, the parade ground, flagpole and several of the buildings at this permanent camp (Sharlot Hall Museum).

The Promberger family in front of the Mayer Owl. The movie poster for "Untamed Youth" dates the photograph to the summer of 1924 (Bill Promberger).

10. The Depression Years and World War II in Mayer

In the 1920s, this was the Price Gasoline Station. By the time this photograph was taken in 1938, it was owned by Frank Garrett and was called the High Way Garage. The Mayer Hotel is on the left and one of the Mayer Apartment buildings can be seen on the right. There are two "visible" style gas pumps. This building was destroyed in the huge snowstorm of December 1967.

A lone burro strolls down Main Street in Mayer in front of James Harris' house in 1941. Martin's Service Station can be seen in the background on the right. Harris was the postmaster at the time. Photograph by Rosena Promberger Minucci.

50 persons, headquarters, officers' quarters, school building, infirmary, recreation hall and a mess hall. Additional buildings included two garages, a warehouse, supply building, blacksmith shop, oil house, shop, two bath houses, two pit latrines and one flush toilet. In 1936, most of the enrollees in Company 822 were from Arizona, but quite a few were from Oklahoma and Texas, and one man was from Brooklyn, New York. The Annual states that "the work project has consisted of twig blight control, trail construction, telephone line construction, rodent control, dwelling construction and erosion control. The enrollee personnel of the company is to be commended for the amount of work accomplished. The camps are made as pleasant a home for the men as possible. There is plenty of activity to enable a man to improve his education and training or to make other constructive use of his time." An April 1940 article from

the *Prescott Evening Courier* reported that "Mayer people and the boys from the CCC camp, who attended the dance Friday night at the Mayer hall, reported having a good time. They are all looking forward to the next dance April 19."

The *Official Annual 1936 of the Phoenix District, 8th Corps Area* summed up the success of the CCC program as follows: "Those young men who have been fortunate enough to serve in this organization, their parents and the nation—all have ample reason to be proud of the part they have played in the building and of the self-improvement that has come to them through the Civilian Conservation Corps."

As the Great Depression wore on, life was continuing in Mayer. Bill Promberger, who grew up in Mayer, lived with his mother in the Mayer household, where she helped care for Sarah Mayer and later Mamie Mayer. The Promberger family had lost their livelihood, and most everything else, when their business, the Mayer Owl, burned in 1931. In May 1941, shortly before Bill graduated from the Mayer High School with the Class of 1941, his sister, Rosena Promberger Minucci, took snapshots of Mayer for Bill. Bill was leaving Mayer almost immediately after graduation for the United States military. Rosena included the church, school, Business Block, Mayer Hotel, White House Hotel and a few overviews. But she also took photographs of the homes and small businesses in Mayer—a little slice of what Mayer looked like in 1941. Mayer today isn't too different from the way much of Mayer looked in 1941, although the cars are newer and the trees are bigger, or have disappeared, and there are some newer buildings. Some of the buildings Rosena photographed in 1941 no longer exist, but her photographs give a good idea of life in Mayer at the end of the Great Depression and the beginning of World War II.

A report entitled "Arizona Lode Gold Mines and Gold Mining," published in 1934, noted that "the Depression, which has thrown many men out of work, and the

The bridge over Big Bug Creek on the Black Canyon Road shows some signs of encounters with automobiles in 1941. Many of the huge cottonwood trees which line the creek are still there today. Photograph by Rosena Promberger Minucci.

Bill Promberger and his mother were living at the Mayer Station, the home of Sadie and Mamie Mayer, when Bill graduated from the Mayer High School in 1941. Carrie Promberger was taking care of Sadie Mayer. The Mayer home looks its age—60 years—in this snapshot taken by Rosena Promberger Minucci in 1941.

A close-up of the old Mayer Mercantile in the Mayer Business Block in a photograph taken by Rosena Promberger Minucci in 1941 shows that Joe Mayer's trees have matured and the hitching rails are still in front of the building. A child and a dog repeat the scenes of earlier times in downtown Mayer.

A photograph taken by Rosena Promberger Minucci in 1941 is labeled "Upper Mayer." The photograph was taken from one of the hills on the south end of town looking northwest shows the back of the Catholic Church (far right) on South Jefferson with the Red Brick Schoolhouse in the distance to the left. The Bradshaw Mountains form the backdrop.

Rosena Promberger Minucci, the photographer who took many snapshots of Mayer in 1941 for her brother Bill, in Mayer in 1946.

One of the Mayer homes photographed by Rosena Promberger Minucci in 1941 is labeled "Kaufman's." It is located at 10070 4th Street.

A family and their dog at their camp in the lower elevations of Yavapai County in 1937 are typical of mining families who were trying to hold onto their claims during the Depression. They may have been camped many miles from other neighbors or from town (Sharlot Hall Museum).

great increase in the price of gold have created an interest in prospecting for and mining gold such as has not existed for many years. Over twenty-six hundred persons are now prospecting for placer or lode deposits of gold or operating on such deposits in Arizona, where less than four hundred persons were employed in gold mines in the state in 1929 and only a few score prospectors for gold were then in the field." During the Depression, some residents of Mayer who had mining claims moved to Prescott or Wickenburg for salaried work; but they continued mining and strove to hang onto their claims and to eke out a living with their mining claims, expecting each day to strike it rich. Often they moved their families back to their claims in the summer, as the Luther Wilson family did when they went back to their camp in Crook Canyon near Mayer. Most spent their lives trying to make their claims produce.

Once the CCC program ended, the Mayer camp stood idle, but not for long. Following the attack on Pearl Harbor by the Japanese on December 7, 1941, the United States government forcefully removed Japanese Americans from "prohibited areas" near the country's borders and relocated them to hastily provided internment camps further inland. One of those camps was the former CCC Camp F-33A at Mayer. The *Prescott Evening Courier* ran a series of brief articles between May 4 and June 5, 1942, on the internment of Japanese Americans at Mayer. The first article reported that

Two billboards on East Gurley Street in downtown Prescott tell the stories of wartime in April 1942. One billboard encourages the purchase of United States Defense Bonds and the other advertises the movie They Died With Their Boots On, *starring Olivia DeHavilland and Errol Flynn. The movie would have played at the Elks Theatre across the street. Photograph by Frank E. Brown (Sharlot Hall Museum).*

about 400 "alien and non-alien Japanese would be removed from their southern Arizona homes to reception centers near Mayer and Cave Creek." A subsequent report stated that a group of 240 "will remain at Mayer until arrangements are made to relocate them to non-prohibited areas selected by the war relocation board." The internees arrived on May 8. On May 16, the *Prescott Evening Courier* reported that 224 American Japanese, 90 percent of whom were American born and many of them natives of Arizona, were housed at the "Mayer Evacuation Center." The article further reported that the "evacuees" were busy with church services and expanding recreation facilities and had "settled down to their center schedule agreeably." Although the Japanese were told that Mayer was a temporary facility, many hoped that it would be permanent. They started vegetable gardens on the former farmland where the camp was located. Former internee Hiro Nomura recalled in an interview in the April 16, 1999, issue of the *Daily Courier* that they could not see the town of Mayer from the camp and they were never permitted to enter the town. But the camp was to be *very* temporary. The *Prescott Evening Courier* reported on June 5 that "the entire population of Japanese who were assembled here early last month has been transferred to the new center at

10. The Depression Years and World War II in Mayer

"Hazel," seated on her motorcycle, is typical of the women who found jobs and social change during World War II as the men went to war and the women became the breadwinners, machinists, farmers, truck drivers, motorcyclists, fliers and heads of families for the duration of the war.

Parker." They were in Mayer less than a month. Today, there are virtually no remnants of the camp to be seen. Arizona Highway 69 bisects the location of the camp and post–1954 construction along the highway has wiped out any remnants of the Mayer Assembly Center.

World War II had a profound effect on all of the United States. A whole generation of young Arizona men, many of whom had grown from boys to men in the CCC camps, went off to war. Young women entered the work force as "the margin for victory" in untraditional jobs that had never before been open to them. They learned to drive—cars, trucks, ambulances and school buses; and, they learned to fly. They joined the Citizens Defense Corp, the Red Cross, the WAVES, the WACs, the WASPs and the SPARs. They suffered the difficulties of rationing, traded coupons and clothing, bartered whatever they could and kept the households and family businesses running. They worked side-by-side with the men who had not gone to war and with people of all races and ethnicity. They filled the labor shortages of the war. They wore pants. Their lives would never be the same again. As Roger E. Kelly wrote in an article entitled "America's World War II Home Front Heritage" in *CRM: The Journal of Heritage Stewardship*, "Much has been written and spoken about how the United States participated in, and was changed by, the world conflict. The nation's home front was like a goldsmith's crucible, recasting relationships between the country's majority and minority peoples into new images and unexpected forms."

Marshall Trimble wrote the Introduction to *Arizona Goes to War*. He wrote:

The "homefront" in Mayer didn't change much during World War II. Many men, like Bill Promberger, joined the military, saw a world beyond their hometowns, served during the war and never moved back "home." This aerial view of Mayer was taken February 26, 1943. It shows the intersection of the Black Canyon Highway and Main Street, which curves off to the right at the bottom of the photograph. From left to right, top to bottom, is Big Bug Creek; the Mayer Station with its white picket fence; the post office; the Mayer Business Block on the corner with the car in front; on the opposite side of the street, the two Mayer Apartment buildings are on the left; next is the High Way Garage; the Mayer Hotel and the Mayer State Bank. The corner to the right of the Mayer Business Block where the Mayer Owl was is vacant.

When the war began, the population of the state was less than half a million. Phoenix was a dusty cow town with a population of 65,414 residents. The metropolitan area, including Mesa, Glendale, Tempe and Scottsdale, boasted 186,000. Tucson had 36,818. Almost overnight, military bases sprang up all over the state.

The economy in Arizona changed dramatically during the war, bringing an end to the Great Depression. Everyone had a job, and people were putting money in the bank again. Cotton and copper returned to boom times. Farmers doubled the amount of land they planted in long-staple Pima cotton. The copper mines operated in high gear. Cattlemen found an eager market for their livestock. Huge government contracts gave rise to the manufacturing industry.

In *Arizona's War Town*, John S. Westerlund set some of the scene for World War II in Arizona. He wrote:

10. The Depression Years and World War II in Mayer   213

Company M, 158th Infantry of the Arizona National Guard poses in the bleachers at the City Ball Park in Prescott on September 24, 1940, at the beginning of World War II. These local men would scatter to many places before they could return to their homes in Yavapai County (Sharlot Hall Museum).

A 1941 snapshot looking east on the Black Canyon Highway in downtown Mayer illustrates that not much has changed since the 1920s. The streetscape is still dominated by the P & E water tank and the Mayer Depot. On the right are Frank Garrett's High Way Garage and the Mayer Hotel. Main Street curves off to the right and a truck which appears to be from the 1920s heads that way. Photograph by Rosena Promberger Minucci.

World War II turned Arizona into an armed camp. At Luke Airfield west of Phoenix, home of the world's largest fighter pilot school, more than 13, 500 pilots trained, as did more than 10,000 combat air crews at Tucson's Davis-Monthan Field. Williams Field, east of Chandler, and Thunderbird II, north of Scottsdale, also trained pilots and cadets. The Army Air Corp established new bases at Kingman, Marana, Douglas and Yuma. Camps Bouse, Horn and Hyder helped prepare soldiers at Maj. Gen. George S. Patton's Desert training Center. Prisoners of war were scattered at two dozen camps across the state, and Japanese Americans were interned at Poston, Sacaton and Leupp. Other camps, fields, and war production camps were built. Del Webb expanded Fort Huachuca, and army engineers built an ordnance depot west of Flagstaff.

In 1942, government subsidies of new production and higher copper prices enabled small mines to make a profit. Within three months of the bombing of Pearl Harbor, over 300 Arizona mines were working twenty-four hours a day, seven days a week.

In *Arizona Goes to War,* Trimble further wrote:

> When the war ended, many of the young men who had trained in Arizona decided to return with their families and take up residency. Footloose veterans learned they didn't have to live where they were born. Arizona was young and vibrant, a good place to raise a family.

By 1950, Arizona's population had grown to 750,000 residents. All of the ingredients were in place to transport Arizona into the modern period of its illustrious history. But Mayer would remain a sleepy little town for some time yet.

11

The Later Years: Post–World War II

On May 8, 1945, Germany surrendered. It was the long-awaited V-E Day in Europe. But World War II wasn't over, as the war still raged on in the Pacific Theater. Arizona Governor Sidney P. Osborn set a solemn tone for V-E Day, stating: "This is a day of thanksgiving and rejoicing. But our rejoicing is tempered by the knowledge that the road to peace is yet long; that American soldiers at this moment are enduring all of the war's hardships and giving their lives in battle in the Pacific areas." It wasn't until August 14, 1945, that Japan surrendered. The surrender released all of the pent-up emotions about the war which had begun on the day Pearl Harbor was bombed—December 7, 1941—the day that will live in infamy. Some doubted the news of surrender, but celebrations nevertheless erupted noisily and spontaneously all over Arizona, from the tiniest towns to the cities of Phoenix, Tucson and Flagstaff. They gradually built to a crescendo as people began to believe that the news of the surrender of Japan was true and that V-J Day was a reality. In *Arizona Goes to War*, Carol Osman Brown wrote of the celebrations in Arizona's small towns: "People in Arizona's rural towns also looked forward to better times and celebrated in their own distinctive styles. Polly Rosenbaum, who served 46 years in the Arizona state legislature, representing Gila County, remembers celebrating the end of the war in the small mining town of Hayden. 'We knew something big had happened because all of the fire bells and whistles from the mines and trucks hauling ore went on and didn't stop. No one except mining officials had telephones, so we all came out of our homes and hollered to our neighbors across the fences as we walked down the hill. Everybody gravitated toward the post office where someone made an announcement, and we all started hugging and kissing. It was a great relief to know that the war was over and we could start living again. It was as if everything had been put on hold.'"

Soon, a huge influx of U. S. servicemen and women descended on Arizona. Trained in Arizona or employed in defense industries during the war, they remembered the sunshine, clear, dry air, open spaces and welcoming western ways. They saw opportunities in this state which had so recently evolved from its Wild West roots into a progressive, modern state on the move. And with the opportunities afforded by the G. I. Bill, which had been co-sponsored by Arizona's Senator Ernest W. McFarland, veterans had an opportunity to go to school, and they took advantage of it. They flocked to Arizona and to Arizona's college campuses in 1945 and 1946. Carol Osman Brown, in *Arizona Goes to War*, wrote of a young female student at Phoenix College. The student, Emma Lou Philabaum, said that Phoenix College "was a small school,

Felicite Roalsted (left), her mother and her step-sister, Bernice Jane Koogler, pause at the side of the road to Mayer on a trip along the Black Canyon Highway on January 31, 1946. With World War II over, tires and gasoline were available and road trips were an exciting change for those weary of the restrictions of the war (Felicite E. Forest).

This snapshot is labeled "first Greyhound to Mayer" but is not dated. The bus drivers and others pose in front of the bus on the Black Canyon Highway in front of the Mayer Business Block. Bus service to and from Mayer must have been a big step for Mayerites shortly after World War II. Greyhound had a bus terminal in Prescott in the 1950s.

sort of a glorified high school, with strict rules and mostly female students during the war. If a student got married while she was in college, she had to get permission from the school president or risk being expelled. Well, that all changed when those veterans came back! It was kind of funny to watch everything being turned topsy-turvy. These men had seen the world and risked their lives. They questioned everything, from rules against marriage and smoking to the high food prices in the cafeteria. Things started changing fast."

In 1947, Arizona's economic engines were still cotton, citrus, copper and cattle. However, as Arizona shifted from an economy based on fighting the war to an economy geared toward developing Arizona, these traditional economic drivers would soon be overtaken by new industries. Tourism (the 5th "C" in Arizona's state seal was "climate") was moving to the forefront, since gasoline and tires were again available and people wanted adventure after the restrictive, lean and stressful years of the war. The manufacturing businesses that had developed during the war were soon to be the leading economic factors in the state. Arizona was becoming the pacesetter for the nation, and thanks to the invention of the evaporative cooler, a pleasant place to live all year.

Meanwhile, sleepy little Mayer was not experiencing the population growth that was flourishing in Maricopa, and Pima counties, and, to some extent, in Prescott. Somewhat off the beaten path, and soon to be bypassed by Arizona Highway 69, Mayer didn't change much in the eight years after World War II. The mines in the area were still operating. Ranching, which was not as volatile as the mining industry, was still the mainstay of the local economy. The railroad still ran through town.

This double header in Mayer includes one of the 1904 Alco-Brooks 2-8-0 locomotives from the old Bradshaw Mountain Railroad, No. 53. This was the "Pollywog" run in the winter of 1946/47 and would be one of the last trips for No. 53 as it was scrapped in 1947. The photograph was taken from in front of the Mayer Station with the Mayer School in the background.

Bill Promberger wrote of this photograph: "This picture was taken in Mayer in the winter of 1946/47. I was working as a brakeman on the Mayer run called the 'Pollywog.' On this day the crew was going to eat on arrival at Mayer. I dropped off the train on arrival at Mayer and had lunch with my Mother at the house in the background. This was the old house that belonged to the Mayer family.... This is the house I lived in from 1934 until I graduated from high school in 1941."

"Downtown" Mayer was still the gathering place for Mayerites and the place to find out what was going on in town. The "old" post office next door to the Mayer Business Block was still the social center of the community. The Mayer Hotel and the White House Hotel were still open. Travelers from the warmer climates to the south appreciated Mayer's cooler weather and slower pace. Some came to visit, some came back again and some stayed. For people traveling from Phoenix to Prescott on the Black Canyon Highway, which ran right through downtown Mayer, Mayer was a convenient place to stop for a few minutes. But that was all about to change.

An article by Ralph Mahoney in the "Days and Ways" section of the *Arizona Republic*, dated May 2, 1954, set the scene for the changes Mayer was about to experience—some good, some not so good. The sub-headline told the story: "Motorists Sigh With Relief as New Road Nears Reality, But it May Spell Doom to a Few Historical Old Towns":

11. The Later Years 219

The Mayer Hotel still graced the corner of Black Canyon Highway and Main Street in the 1950s as an apartment building and, as usual, the porch was getting some use. A more modern gas pump had replaced the old "visible" style pumps at the High Way Garage and the lighted sign advertised Chevron Gasoline (Herb and Dorothy McLaughlin Collection, Arizona State University Libraries).

Time and asphalt are changing the complexion of the 95 miles of road between Phoenix and Prescott, and predictions have been made that Arizonans now in their 60s may live to see the Black Canyon Highway paved from one end to the other.

Highway department officials hazard a guess that the Prescott leg will be paved within two years—if sufficient money is appropriated.

Business men of Mayer, Humboldt and Dewey won't share the happiness of the motoring public. They're on the fringe of the highway and they'll be hard hit if their gasoline station, grocery store, restaurant or tavern depends entirely on the tourist trade. A few will move the half-mile or mile to the highway; others will fold if they can't entice the motorist off what will then be the paved and beaten path.

Despite what obviously will be a serious situation, pessimists can be counted on the fingers of one hand.

Mayer has existed for over three-quarters of a century—which is a long time in Arizona's youthful history—and road or no road, it ought to be good for a few more years. Humboldt and Dewey are both in their 50s.

With its history behind it, Mayer has become a sleepy little town of 500, comfortably reclining on a hillside, watching the world go by.

The Black Canyon Highway has spread out much farther than the 10 miles of Black

In the 1940s and 1950s, Ray's Market occupied the Mayer Business Block. The market and Holsum Bread were advertised on the side of the building. Noble Ray Crawford owned Ray's Market. Most of the signs are faded but still legible today.

Canyon (at Bumble Bee) for which it was named. Its importance as a stage route between Phoenix and Prescott in the early days parallels its importance as a high-speed shortcut to the north country today. Undoubtedly, it will gain more importance as the years go by.

A Mayer Chamber of Commerce brochure from the mid-1950s touts Mayer as having "Arizona's Finest Year Round Climate" and as an "Ideal Retirement Town." Regarding the climate, the brochure stated: "For full year round living, Mayer and vicinity has one of the most ideal climates in the state. Located at what can be considered the snow line, near where the mountains meet the desert, the winters are tempered by the desert sun and the summers are cooled by the mountain breezes.... The elevation of 4,341 feet above sea level gives Mayer a fine year round climate. If you are retired or about to retire or if you have asthma or other respiratory ills or arthritis requiring a dry climate, you can find it in Mayer or nearby. If you like the COUNTRY life, then the little town where there are no strangers will certainly interest you." And as to the appeal to retirees, one retired resident who had come to Arizona for his health remarked that "We have settled here and feel that we have found the best spot in Arizona." Regarding property values, the brochure provided the following information: "At present a nice lot large enough for a two bedroom retirement home can be had at prices between $650.00 and $1,000.00 each with water, electricity and natural gas available. Mayer has no industry but does have two old fashioned hotels, one motel, one restaurant, with another planned for this spring, two trailer parks, two grocery stores, two taverns, one antique and rock shop, two churches, Texaco, Standard and Richfield service stations, three real estate brokers and three subdivisions with more in the planning stages." This presents a pretty accurate picture of Mayer as businesses along the new Highway 69 began to develop (motel, restaurants) after the re-routing of the Black Canyon Highway away from downtown Mayer.

In the spring of 1959, the *Yavapai County Messenger* reported on the "doings" at Mayer, featuring on the front page a new home being constructed in Mayer, noting "Elden Sears is developing the 25 acre subdivision and already has sold four lots. Sears ... says he expects Mayer to become a favorite spot for retired people since Mayer had the 'best climate in Arizona.'" In a full-page spread in the same issue under a headline of "Mayer Isn't Yet—And Refuses to Be—A Ghost Town," the columnist reports that Mayer is "slowly getting up steam after period of quiet when the mines closed." The columnist further wrote:

> Back in the early 1900s there was a saying that "All roads lead to Mayer." That was when Mayer was a bustling, thriving mining center for the numerous mines which dotted the nearby Bradshaw Mountains.
>
> After a period of standing still, the town is once again humming to the tune of the hammer, saw and cement mixer.

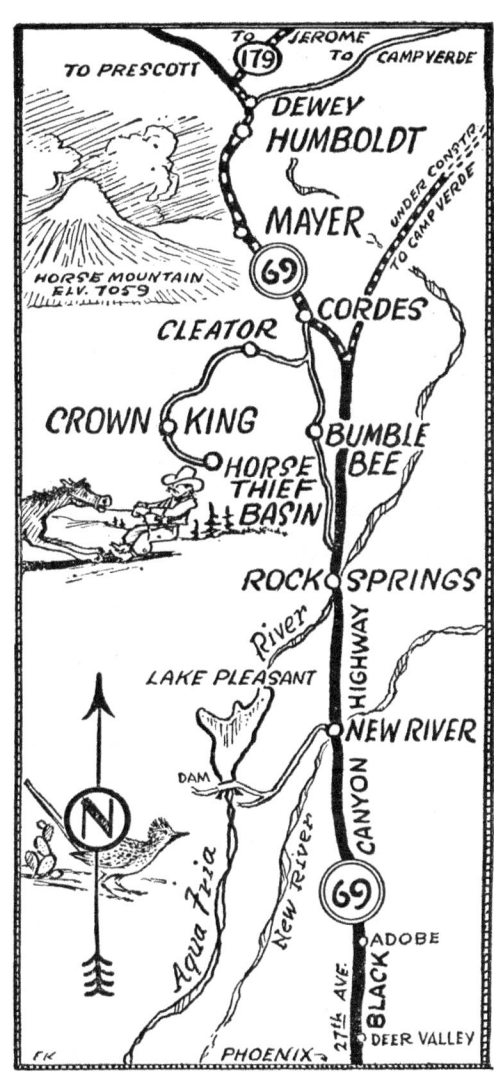

A 1954 map shows the start of what is now Interstate 17 at the north end of 27th Avenue in Phoenix. On the map, Interstate 17 is still called the Black Canyon Highway but is also designated as Highway 69. Today, the modern Highway 69 starts at the unlabeled "Y" on the map at what is now Cordes Junction and goes through Mayer, Dewey-Humboldt, Prescott Valley and ends at the intersection of Highways 69 and 89 in Prescott. The unlabeled road "to Camp Verde" is now Highway 169. And, indeed, they are all paved. Map from the **Arizona Republic,** *May 2, 1954, "Arizona Days and Ways," article by Ralph Mahoney. Used with permission. Permission does not imply endorsement.*

In 1959, the Black Canyon Highway led from Phoenix through the Black Canyon and on through the grasslands and the mountains to Prescott. Today, only the stretch from Cordes Junction to Prescott is considered to be the "Old Black Canyon Highway" and it is a four-lane or more, partially divided State Highway which bears little resemblance to this photograph of the Black Canyon Highway in 1959 (Sharlot Hall Museum).

Still in the "country" in the 1950s, the Black Canyon Highway leaving Mayer to the north crossed Big Bug Creek between the giant Fremont Cottonwood trees that lined the creek. Some of these trees are still there today. The Mayer Station is on the right just a few hundred feet closer to downtown Mayer.

This 1952–53 photograph is labeled, "Big doings on the streets of Mayer in front of Ray's Market." The child hanging onto the hitching rail, as children had done for more than forty years, is Tom Crawford, son of Ray's Market owner Noble Ray Crawford.

Emil Mueller's Service Station has seen many businesses come and go, but the distinctive style of the building is a give-away as to its original purpose. Mueller stayed put on Main Street when many other businesses were moving to the new Highway 69. Photograph by Nancy Burgess, 2007.

This 1968 view of Highway 69 in Mayer looking northwest toward Prescott includes a vacant lot on the corner of Central Avenue and Highway 69 where the Circle K is now located. The area on the same side of the highway to the northwest between the highway and Big Bug Creek was reportedly the location of the CCC Camp. The Catholic Church is visible on the hill to the left and the Red Brick Schoolhouse is on a hill to the right. The onyx quarries are to the right of the school in the flat areas between the highway and the creek (Sharlot Hall Museum).

MAYER WATER COMPANY

"Congratulations to Seniors"

Box 96   632-7744

Mayer, Ariz.

The Mayer Water Company informally dates to the days of Joe Mayer when he brought water from Grapevine Springs to Mayer. The privately owned Mayer Water Company is still in business today. An advertisement from the Mayer High School yearbook congratulates the graduating seniors of the Class of 1964.

11. The Later Years

In 1955, Mayer was just beginning to wake up from a lengthy slumber. Although new houses and businesses would soon be going up, the Black Canyon Road into Mayer from the northwest still presented an image of a sleepy, off-the-beaten-track town. A car is heading into Mayer while a steer is heading out of town. Neither appears to be in a hurry (Sharlot Hall Museum).

Except for some new paint, not much changed for the Mayer Business Block through the 1980s. This photograph shows the building along with the old post office building shortly before the properties were sold out of the Mayer family. Photograph by Nancy Burgess, 1989.

Yavapai County experienced one of the heaviest and longest-lasting snowstorms in recorded history in December 1967 and January 1968. Although Mayer does get snow, due to its lower elevation it is usually light and does not last long. However, the record-breaking "Snow of 1967" took the awning down on the Mayer Business Block and destroyed the High Way Garage across the street. The awning was never replaced.

    The "old Town" is a block long row of buildings now closed and fading. The old hotel has long since registered its last guest and the pot bellied stove sits idle and cold in the old lobby amidst faded chairs and sofas. The once modern bank has closed its doors after being used as a malt shop for awhile. Across the street the barber shop, general store and tavern have been locked. The long veranda and iron railings with hitching rings still remind a visitor that this was once the town of Mayer.

    Just beyond this area is the homestead of Joe Mayer, who founded the town in 1880. Mayer's eldest daughter, Mamie B. Mayer, still lives there, perhaps reflecting on memories of once-glorious days of the town. It is a place where cattle roam the deserted streets and a foot-loose burro finds shade alongside the deserted railroad station.

    However, if a person looks up to a high bluff overlooking the town he will see that Mayer is again striving to keep from being another ghost town. Elden Sears, a man with an eye for the future and a love for Mayer is subdividing a 25 acre section atop the bluff and erecting modern homes.

    At the foot of the bluff alongside busy Black Canyon Highway, a person can find other continued signs of continued growth. A new service station, general store, tavern, treasure shop and six unit motel have been built for the comfort and service of the traveling motorist.

    After talking with residents of Mayer, you realize they are proud of their past, but also looking forward to a bright new future.

Built in the late 1950s on the new Highway 69, Duncan's Oak Hills Motel and Restaurant took advantage of the highway location and the modern travelers who drove the new "scenic" highway. The restaurant became a comfortable gathering place for Mayerites to socialize and have a good meal. Photograph by Tom Reed, postcard published by Phoenix Specialty Advertising Company, Phoenix, Arizona.

The late 1950s and the 1960s were probably the low point for Mayer. According to Yavapai County Chamber of Commerce, the estimated population of Mayer in 1962 was 350. At the time, the Mayer Elementary School had 62 students, and the junior-senior high school had an enrollment of 32. The onyx quarry was still in business and there were seven retail establishments. Cattle ranching was the major agricultural business in the area. Although the population was growing slowly in the 1960s, by 1970, the estimated population of Mayer had doubled to 700. Quite a few new residents were moving in, but the popularity of Mayer as a "commuter" community did not really come about until the late 1970s and the 1980s. There were few businesses in downtown Mayer in the 1960s and 1970s, as most everyone had moved to the highway, just as Ralph Mahoney predicted in his May 1954 "Days and Ways" column of the *Arizona Republic* when he stated that businesses would "fold if they can't entice the motorist off what will then be the paved and beaten path."

In an interview for the *Prescott Courier* in 1982, Irene Thompson McDonald, who was born in Prescott in 1905 but grew up in Mayer, said that "where Duncan's Restaurant (Oak Hills) is now there was a ranch and we lived on it. The Mayers had homesteaded it and my father rented it." This was probably part of the original 160 acres that Joe Mayer bought from William Muncey in 1882.

During the mid-1960s, a complementary weekly newsletter, the *Desert View* was published by Mayerite Dwight L. Bockman. It was strictly a regional publication, covering Mayer, Black Canyon City, Cleator, Cordes Junction, Humboldt and New

A current view of the Duncan's Oak Hills Motel and Restaurant shows that it hasn't changed too much in the last fifty years. It is now located on a frontage road along Highway 69, which is now a divided highway through Mayer. Photograph by Nancy Burgess, 2011.

Emil Mueller, who, like quite a few Mayer residents, came from Switzerland, lived close to his service station on Main Street. His home was typical of many of the houses in Mayer built after World War I. Photograph by Rosena Promberger Minucci, 1941.

River. A few of the businesses advertised in the *Desert View* were from Prescott or Phoenix, but most were local. The advertisements of the time clearly show the need on the part of the small town businesses to be as diverse as possible. The newsletter featured poetry, a little philosophy, new buildings, new businesses and the local "news" of who was visiting in town, births, bake sales, property purchases, parties, the catch from a fishing trip, meetings, directories, accidents, weather, want ads and funeral notices and snippets of local history along with historic photographs. Mr. Bockman wrote a weekly column entitled "My View." One of the businesses featured in the December 16, 1966, issue was "Laura's Café" on Highway 69 at the south end of Mayer. The restaurant was opened in December 1966 by

Emil's TEXACO SERVICE

- Auto Parts
- Hardware
- Electric Supplies
- Ammunition
- TV & Radio Tubes
- Hunting & Fishing License

Main St. and Wicks Ave.
Phone 632-7791   Mayer

Vaughan's Service & Repair

☆ Auto Refrigeration
☆ Appliances
☆ Electrical Work
☆ Plumbing

BUY - SELL - TRADE
SECOND-HAND GOODS

OLD POST OFFICE BLDG.
MAYER 632-7470

As Mayer changed with the times, Emil kept his gas station, but branched out as much as he could to keep his business going. This advertisement from the Desert View *in 1966 shows that he provided quite a few goods and services to Mayer, including hunting and fishing licenses. A fisherman would have to go quite a distance from Mayer to do any fishing.*

Another long-time Mayer business was Vaughn's. When the post office moved out of the Mayer's building in 1958, Vaughn's must have moved in shortly thereafter. The business had a variety of merchandise and by the time this advertisement was published in the Desert View *in 1966, was offering plumbing and electrical work in addition to second-hand merchandise.*

There has almost always been a grocery store in Mayer, starting with Joe Mayer's store in the Mayer Station in the 1880s. In 1964, the "Hitching Post" offered staple and "fancy" groceries.

In the 1970s, Duncan's Oak Hills Motel and Restaurant was a busy place, specializing, according to the postcard, in "good food and quiet atmosphere." Highway 69 runs along the lower right corner of the photograph. Photograph by Charles P. Kendall, postcard published by Kendall Photography, Yuma, Arizona.

In 1989, the Mayer Business Block and the Mayer Apartments were listed in the National Register of Historic Places by the United States Department of the Interior. Typical of late 19th century brick commercial buildings, they were not distinctive architecturally and, at the time, were not in very good condition. However, changes to the buildings were minimal and their historical importance to the community of Mayer was based on their association with Joe and Sadie Mayer. Photograph by Nancy Burgess, 1989.

Laura Johnston and Ethel Scott. Their advertisement invited truckers for their specialty, Spanish food. Emil Mueller was from Switzerland, as were quite a few citizens of Mayer in the early 20th century, and came to Mayer in 1920 after World War I. He opened his first "service garage" in Mayer in 1934. A 1966 Mayer advertisement showed that Mueller, who had two gas stations on Main Street in Mayer, had stayed put at his original locations *off* the new highway, offering auto parts, hardware, electric supplies, ammunition, TV and radio tubes and hunting and fishing licenses at his Texaco service station. Vaughn's Service and Repair, also known as Vaughn's Bargain House, housed in the "old" post office next door to the Mayer Business Block, offered auto refrigeration, appliances, electrical work, plumbing and second-hand goods. Mrs. Coghan, across from the Mayer Jail, advertised "Tupperware." The Mayer Enco service station, run by Orin Culver, carried groceries and did tire repair. The Mayer stone quarries (Givens Stone Yard) were still in business, offering building stone, Mayer onyx, fireplace stone and "custom cuts" on a 24 inch saw, and, just for variety, Saanen milk goats. Bolding's Market, on the west side of Highway 69, was featured in *Desert View* in November 1966 as a "fine, modern Associated Grocer facility bringing to this small community just about any service to be found in the big city." Thanksgiving turkeys were 41 cents a pound, Mrs. Cubbison's turkey dressing was 59 cents and ice cream was 25 cents a pint. Just down the street were the Duncan's Oak Hills Motel and Restaurant, which advertised "good food and quiet atmosphere."

The 1980 population of Mayer was around 1,800. Sixty percent of the population were retirees. Those who were employed worked in local businesses or in Prescott or Phoenix. By the late 1970s and in the 1980s, the Mayer Business Block was occupied by the Mayer Elder's Club and other tenants, and downtown was showing some true "signs of life." In July of 1989 the Mayer Business Block and the Mayer Apartments were listed in the National Register of Historic Places. One of the criteria for listing the buildings was their association with a person who is important in the community: Joe Mayer. Other factors which made the buildings eligible for listing in the National Register were the age of the buildings and their "historic integrity"; that is, how original they were in construction and appearance to the day they were constructed. Also around the same time, the buildings which had been in the ownership of the Mayer family for decades were sold to other owners. This included the site of the Mayer Station, the Mayer Business Block, the "old" post office, and the Mayer Apartments, plus some vacant land, along with numerous mining claims.

But new residences, new residents and businesses new and old were not the only things going on in Mayer. Yavapai County has a long tradition of moviemaking, starting in 1912. In some ways, the landscape of Arizona and Yavapai County have helped shape the world's romanticized perception of the Old West as portrayed by Hollywood. Early Westerns depicted a sense of morality, freedom and rugged individuality. As Jay Boyer, Ph.D., wrote in a chapbook, *The Big Screen and The Big Sky*, written to accompany a film series at the Coconino Center for the Arts in Flagstaff, Arizona: "The western is perhaps the most purely American of any of our film genres. It is, by definition, set in the American *West*, generally the Southwest, and we are to understand from this that we are looking at the last frontiers of the United States to be settled and civilized." Boyer also wrote: "There are two basic locales for the Western. The

A William Fox Studio movie crew gathers in Yavapai County for a movie shoot sometime in the 1920s. One woman (far right) joins the group. She is probably the "star" of the movie. The clothing helps differentiate the film crew, drivers, producers, etc. from the cast (Jeffrey Ogg).

Tom Mix, who came to Prescott in 1913 as a stuntman for the Selig Company soon became a movie star. Mix made dozens of movies in Yavapai County. Here, he is on a set in the Granite Dells outside Prescott, one of his favorite places to film. Mix is the man third from left in the "ten gallon" hat (Jeffery Ogg).

When scouting for a location for a television series pilot led the producers to Mayer in 1993, they "liked the looks of it." The Mayer Business Block was made to look even more run-down than it did at the time (they should have come in the 1980s). Photograph by Nancy Burgess, 1998.

A proposed television series was called **Harts of the West.** *Mayer was supposed to be "Showlow," Nevada. The locals had a great time with the process and had their pictures taken next to the Mayer Business Block with the personalized chairs for the producer and cast. The photograph was taken in 1993 and the people are unidentified.*

One of the floats in the May 2011 Mayer Daze Parade was provided by the Desert Heritage Museum, Inc. The float is headed southeast on Central Avenue (the old Black Canyon Road) and has just passed the Mayer Business Block. Photograph by Nancy Burgess, 2011.

first is the frontier itself.... The second locale of the Western is what we might call the *self-contained settlement*. I'm thinking about the cavalry fort, the ranch, the Western town, and the like.... If it is a town, then the settlement is sure to have one street, the main street, one saloon, one hotel, one doctor, one lawyer, one newspaper, one stable, one of almost everything we associate with civilized life—and most important in this regard, one sheriff." This description of the quintessential Western movie town would have generally fit many of the towns in Yavapai County throughout the late 19th and early 20th centuries, including Mayer.

As Dewey E. Born wrote in an article for the *Prescott Courier* in July 1999, entitled "Tom Mix Comes to Prescott to Make Movies," "In 1913 the Selig Company came to Prescott with ... a stuntman named Tom Mix. Tom Mix had been a cowboy, ranch foreman, sheriff, deputy U. S. Marshall, Texas Ranger and rodeo performer before joining Selig in 1909. When they started making movies in Prescott, he quickly moved from stunt man to star." Tom Mix made movies around the Prescott area for many years, first with Selig, then with the William Fox Company, and later with his own company. He owned a ranch which was located where the current Prescott subdivision of Yavapai Hills is today. Local lore says that Mix slept in the big brass bed in the White House Hotel in Mayer. He is credited with starring in 336 feature films. He produced 88 of them, wrote 71 and directed 117. His earliest films date from 1909, while his last film, *The Miracle Rider*, was made in 1935. Mix was the Western movie hero of the 1920s and 1930s. Anyone who is a fan of the old Westerns is a fan of Tom Mix. He made so many movies in the Prescott area that the scenery became famous and led to many other films, movies, television programs and commercials made in Yavapai County and Prescott.

In 1993, the CBS television pilot for *Harts of the West* was filmed in Mayer and, for a short while, Mayer became "Showlow, Nevada." An April, 7, 1993, issue of the *Prescott Sun* ran a front page article about the filming of the pilot, which starred Beau Bridges and Harley Jane Kozak as Dave and Alison Hart and Nathan Watt and Meghann Haldeman as their children. Starring as the crusty old rancher was Lloyd Bridges. The reporter, Stephanie Hanville, wrote that Rob Cook, the owner of the buildings used in the movie (the Mayer Business Block) told her that "they looked in six states for a place like this. They liked the looks of it." The article further stated that Rhonda Cook, Rob's wife, said that "It was really exciting for everyone. Everyone hung out at the post office and watched. They let the kids out of school to watch." Rhonda Cook worked at the local Valley National Bank and when asked if she was involved in the movie, she said that "I had to ride a horse. They wanted a horse so I got a local horse and I rode it up to them on my lunch hour from the bank." Mayerites sat on the front porch of the Mayer Hotel (then an apartment building) and watched. They ended up being filmed for the movie right where they sat. According to locals who watched the goings-on, at first the film crew dawdled around painting and building and, then, all of a sudden, they were done. The local observers described the appearance of the Mayer Business Block after this "work" was completed: "That building across

Just as in the 19th century, a stagecoach heads down the Black Canyon Road in the 2011 Mayer Daze parade. The stagecoach is sponsored by Truly Nolen Company. Photograph by Nancy Burgess, 2011.

Adults, kids and dogs gather at the intersection of Central Avenue, Main Street (to the left) and Oak Street (to the right) as entrants in the 2011 Mayer Daze parade pass the Mayer Hotel and the Mayer Business Block. Photograph by Nancy Burgess, 2011.

Mayerites, who appreciate the mild, four-season climate, rural lifestyle and friendly atmosphere of Mayer and who commute to Phoenix or Prescott, hit Highway 69 for the 25-mile drive to Prescott or the 75-mile drive to Phoenix. Instead of a day or more to make the trips by horse or stagecoach, it only takes a short time to drive the distance. This photograph shows Highway 69 on a beautiful fall day. The vehicle is headed south through Mayer. Photograph by Nancy Burgess, 2011.

the street used to be red and now it's white and that tin roof used to be shiny and now it looks old and rusted…. They put up all the signs and even made the windows look all dirty." Cook mentioned that "it was amazing how fast they changed things, put it up and took it down. It took them 10 days to set it up and three days to film." There were few restrictions on the set, and Mayer residents wandered on and off the set taking pictures and having their pictures taken. A few asked for autographs. Once the filming was finished, everything was put back the way it was, and there was no visible evidence that the film crew had been to Mayer. But friendships were made, and those were not to be forgotten.

One more "big event" was occurring in Mayer on an annual basis: "Mayer Daze." Mayer Daze started as a fund raising barbeque in 1962. Today, the Mayer Daze celebration includes a breakfast at the Mayer Elder's Club, band music, a parade featuring local and regional groups, students, horses, tractors, historic vehicles and bands. The theme and participants vary from year to year, but almost all of Mayer, plus many visitors, turn out for the event and the barbeque following the parade. The 2011 Mayer Daze was held on May 14. The theme was "Out of the West Came Myths and Legends." This celebration is a good opportunity for Mayer to show off its historic downtown and the folks who live and work in Mayer. Traditionally, the proceeds from Mayer Daze go to the Mayer Recreation Center.

By the end of the 1980s, Mayer had settled into an identity as a friendly, quiet, small, rural town with enough amenities to serve the local population on a day-to-day basis. Located close to the Town of Dewey-Humboldt, just 15 miles from Prescott Valley and 25 miles from Prescott, a quick drive on Highway 69 would take a Mayerite to the "big city." Seventy-five miles in the opposite direction leads to Phoenix and the Salt River Valley communities. Today, with 65 and 75 mile-an-hour highways, many commuters have made Mayer their home. They commute to Phoenix, or to Prescott Valley or Prescott. The rural lifestyle, reasonable cost of living and easy access to that highway that had such an impact on the town of Mayer have made Mayer an even more popular place to live.

12

The Modern Years: Mayer Today

Of the more than six million people who live in Arizona as of the 2010 United States Census, 211,000 are in Yavapai County. From 2000 to 2010, the population of the county increased by 26 percent while the population of the state increased during the same time period by approximately 24 percent. Currently, the population of Yavapai County is nearly twice the total population of the entire Territory of Arizona in 1900. The population of Yavapai County was fairly stagnant from 1920 through 1960, hovering around 25,000. But by 1970, increases in population each decade began to climb and the most dramatic change occurred between 1980 and 1990 when the population soared from 68,000 to almost 108,000. With a land area of 8,123 square miles, Yavapai County has a population density of only 26 persons per square

A 1981 overview of Mayer facing southwest shows the original part of the town, which has now spread out far beyond the original townsite. Highway 69 runs through the center of the photograph from right to left with the bridge over Big Bug Creek to the right. The hill on the far left sports the Mayer Smokestack on top and is the site of the Grey Eagle Smelter. The red Brick Schoolhouse is on the far right. Downtown Mayer is in the center of the photograph. Although Mayer has grown quite a bit since 1981, the setting has not changed, and the Bradshaw Mountains, with their secrets of hidden treasures, form the backdrop for the town (Sharlot Hall Museum).

12. The Modern Years 239

Modern amenities come with growth, and Mayer relies on Yavapai County for most of their services, such as the library, which is a member of the Yavapai County Library District, shown here. Photograph by Nancy Burgess, 2011.

The old Black Canyon Highway still runs through Mayer, only today it is known as Central Avenue. Here, the bridge over Big Bug Creek is still flanked by the signature Fremont Cottonwoods. Except for the fact that the road is paved, the scene at this location has not changed much in the last 100 years. Photograph by Nancy Burgess, 2011.

Highway 69 looking south at the intersection with Central Avenue, which leads to downtown Mayer and the historic buildings at its center. The Mayer Hotel, White House Hotel, Mayer State Bank Building, Mayer Business Block and the old post office building are all still an important part of the history and economy of Mayer. Photograph by Nancy Burgess, 2011.

The old post office building owned by the Mayer family has recently been remodeled and houses an antique and collectibles store, which is open for business. Photograph by Nancy Burgess, 2011.

12. The Modern Years 241

New trees and a split rail fence have been added to the front of the Mayer Business Block, but the old hitching rails are still there. The building has been rehabilitated and restored by long-time current owner Mike Connors. Photograph by Nancy Burgess, 2011.

Although it is no longer open, the White House Hotel has been lovingly preserved by Clyde and Sonja Hickey McDonald, including the bunkhouse and barns in the back. Photograph by Nancy Burgess, 2011.

The Mayer Hotel, now housing apartments, has lost its bracketed eaves and upstairs railing but has gained replacement siding. Unfortunately, these changes have made it ineligible for the National Register of Historic Places. But it would still be easily recognized by anyone who frequented the hotel in the days of Joe Mayer. Photograph by Nancy Burgess, 2011.

The former Catholic Church on Jefferson Street, built in about 1908 and which remained a church until 1982, was a teahouse for a time. It has changed quite a bit on the exterior with the addition of a wrap-around porch, but it still recognizable due to the belfry on the roof. It is now a private home. Photograph by Nancy Burgess, 2011.

A view west-northwest from the north side of Highway 69 shows the 1950s development along the highway, including the Oak Hills Restaurant. Photograph by Nancy Burgess, 2011.

A view southwest from the north side of Highway 69 shows the newer development along the highway, at the intersection with Central Avenue including the Dollar Store next to the Oak Hills Motel (to the right). Photograph by Nancy Burgess, 2011.

Burr and Annie Mayer's Bungalow on Main Street was built in 1906 after their first home burned. According to local information, it has coffered ceilings, oak trim, doors and floors and built-in cabinets typical of the Arts and Crafts era when it was constructed. Photograph by Nancy Burgess, 2011.

Mayer's landmark, the smokestack, still stands as a sentry watching over the old town of Mayer. Photograph by Nancy Burgess, 2011.

mile. Incorporated communities include Prescott, Prescott Valley, Chino Valley, Cottonwood, Clarkdale, Camp Verde, and Sedona. The rest of Yavapai County consists essentially of rural, unincorporated towns, although there are a few small incorporated towns, such as Jerome and Dewey-Humboldt.

Mayer is one of those unincorporated communities which relies on the Yavapai County government for its basic services, such as law enforcement, Justice Court, library, road and street maintenance and building and zoning enforcement. The old Black Canyon Road, which runs right through downtown Mayer, is now Central Avenue. Main Street is above (as in uphill) Central Avenue to the southwest. The part of Mayer above Main Street was usually called "Upper Mayer" and the part along the Black Canyon Road (now Central Avenue) was "Lower Mayer." Arizona Highway 69, a four-lane parkway, passes east of old downtown and the old Black Canyon Highway and is maintained by the State of Arizona. Mayer has its own water

A sand and gravel company (G&S Gravel, Inc.) has taken over the old sites of Arizona City and the Boggs Smelter located northwest of Mayer. The site of the Boggs Smelter is off to the left on Big Bug Creek behind the undisturbed hill. Photograph by Nancy Burgess, 2011.

company, a volunteer fire department, an Elder's Club, a Community Center and two cemeteries. The United States Census Bureau recently reported the "Zip Code Business Patterns" for 2009. Statistics for the Mayer area clearly show that it is a part of Yavapai County where small businesses predominate, with most of the 65 Mayer area businesses employing no more than ten people. Most of the businesses serve the local population with retail trade having the largest number of businesses with 15. Leading the service and trade businesses are the construction trades. The Mayer area businesses included manufacturing (trusses, metal work and machining), wholesale electrical equipment, commercial trucking, a bank and an insurance agency, a lawyer, a doctor, a medical lab and home health care services, a real estate management company, artists and food service businesses. One of those food service businesses is the Big Bug Station, which is located in the same space in the Mayer Business Block as Joe Mayer's Restaurant was in 1904. According to an article in April 2011 by Heidi Dahms Foster in the *Courier*, the owners of the business cleaned, polished and refurbished items they found in the building to decorate with and created a new gathering place for the community, a "large, airy, comfortable and historic place where residents and visitors can enjoy good food, coffee and company." The White House Hotel is no longer open and the Mayer Hotel has been apartments for many years. The Mayer Apartments were converted decades ago from three units each to one unit each, and the "moon houses" in the back yard were removed in the 1980s. The Mayer State Bank is now a florist's shop. The original Catholic Church has been converted to a private home after a stint

Heavy equipment operates Monday through Friday taking down the hills in the area of the old mining claims. Recently, elevators sit idle at the sand and gravel company. Photograph by Nancy Burgess, 2011.

Joe Mayer's Onyx Mine has been through a number of ups and downs and openings and closings along with numerous owners. However, today, there is a substantial stockpile along with some large specimens of beautiful onyx in the yard of the current owners, Stoneworld. Photograph by Nancy Burgess, 2011.

as a Victorian Tea House. Out on Highway 69, Mayer has a motel and restaurant, Circle K convenience store, grocery, Dollar Store, and gas stations, along with a number of small businesses. At 129.5 feet in height, the landmark symbol of Mayer, the Great Western Smelter Smokestack, which was built in 1917, still stands on the hilltop southeast of Mayer.

An article in the June 7, 1996, issue of the *Courier* addressed the growth in Mayer, stating that Mayer experienced record growth in 1995, and that "the community 16 miles east of Prescott Valley also has been shining its own beacon, attracting large numbers of Phoenicians seeking a rural and economically-attractive way of life. Some are even commuting from their Mayer homes to jobs in the Valley. A lot of people are moving from the Valley to this area. They are finding the rural atmosphere conducive to raising a family."

Today, there are no categories listed in the United States Census Bureau "Zip Code Business Patterns" for 2009 for ranching or mining in the Mayer zip code—trades that kept Mayer going for many decades. But ranching and mining are still viable contributors to the economy and culture of Mayer. Small mineral claims and mines can be found throughout the area and placer mining is more popular than in the recent past due to the economic conditions of the times. Sand and Gravel operations have taken over the old mining camp of Arizona City, about five miles north of Mayer on Big Bug Creek, and the Boggs Smelter. Although it has had numerous owners, the Onyx Mine is still operating. Many of the ranches of the late 19th and early 20th

Dewey started out as a stop along the P & E. Located along the Agua Fria River, the community is well known for its commercial farming. Young's Farm grew corn, pumpkins and other crops for more than 50 years at the intersection of Highways 69 and 169. Currently, Mortimer Family Farms is farming the same land. Photograph by Nancy Burgess, 2011.

A view along the old Black Canyon Highway in Dewey facing southeast, of the former Young's Farm with the Mingus Mountains in the background. Photograph by Nancy Burgess, 2011.

The mines at Crown King have been closed for a long time and the railroad was abandoned decades ago, but the town of Crown King, with its cool summers, has become a popular tourist destination, especially the historic Crown King General Store. It was built right across the street from the Bradshaw Mountain Railroad Depot, which no longer exists. Photograph by Nancy Burgess, 2010.

Crown King has become a community of mostly summer homes. Although the summer weather is appealing to those who want to get out of the heat of the lower desert, the winters are another story. The road to Crown King, which is the old Bradshaw Mountain Railroad bed, is an adventure unto itself. Photograph by Nancy Burgess, 2010.

centuries are still in business today. Subdivision development has spread throughout Yavapai County, and several of the area ranches which have easy access to Highway 69 have succumbed to modern development, such as Bensch Ranch. However, an aerial view of the Mayer area still shows thousands of acres of open land. The Prescott National Forest, along with United States Bureau of Land Management (BLM) and the State of Arizona make up the ownership of much of Yavapai County, and Mayer is nearly surrounded by public lands. Pockets of private land, along with leased BLM or state land make up most of the ranches in the area. Mining claims are scattered throughout the Mayer area from Black Canyon City to Prescott, along with ghost towns and ghost camps from the days of the mining booms. There is little or nothing left of many of these communities; some have simply been completely removed, and some have just melted into the earth. Old Cordes, Cleator and Bumble Bee hang on. Some of the old camps and towns have evolved into modern communities, such as Breezy Pines at the old site of the mining town of Poland, and Dewey, originally a stop on the Prescott and Eastern Railroad and now a part of the incorporated Town of Dewey-Humboldt, which lies adjacent to the Agua Fria River. Farming is still an important part of Dewey's history and economy and the Old Black Canyon Road runs adjacent to Highway 69 through Dewey. The historic mining town of Humboldt still

The 1902 Mayer School has been moved once and served many uses in the community of Mayer. Today, it is a private home, but its origins as a school and the Ladies' Aid Building are still evident. Photograph by Nancy Burgess, 2011.

There are still enough students in Mayer to "make up a school" and, in fact, the enrollment in the Mayer elementary school outgrew the Red Brick School House quite a few years ago. The new Mayer Elementary School serves the Mayer area. The Mayer High School is in Spring Valley. Photograph by Nancy Burgess, 2011.

has its historic downtown plus a modern commercial area along Highway 69. The site of the largest mine in the Bradshaw Mountains, Crown King, located in a beautiful setting, continues to attract tourists, along with summer and a few year-round residents (you would have to be a dedicated visitor to tackle the old railroad grade of the Bradshaw Mountain Railroad to get to Crown King). Arcosanti, begun on a raw, high desert mesa of the Bradshaw Mountains in 1970 near the intersection of Interstate 17 and Highway 69 at Cordes Junction, is a utopian community designed by architect Paolo Soleri where students live, build and work. Arcosanti is designed according to the concept of "arcology" (architecture + ecology). In arcology, the built and the living interact. When complete, Arcosanti will house 5,000 people, demonstrating ways to improve urban conditions and lessen the destructive impacts of man on the earth. Its large, compact structures and large-scale solar greenhouses will occupy only 25 acres of a 4,060 acre land preserve, keeping the natural countryside in close proximity to urban dwellers. Arcosanti hosts 50,000 tourists a year with tours of the community, a gallery, bakery, and café.

The Mayer School no longer has classes in the Red Brick Schoolhouse. The 1902 Mayer Schoolhouse, later known as the "Ladies' Aid Building," is a private home. The new Mayer School on the opposite side of the creek from the Red Brick Schoolhouse serves grades kindergarten through eighth. School-age children from the neighboring ranches still go to school in Mayer, but instead of staying at the Mayer Station with Joe and Sadie and their children, they can go home every day. High school students in the region have attended the Mayer High School in Spring Valley since 1981. Orme School is still teaching and boarding students from all over the world at the Orme Ranch.

As is the case with most small Western towns, Mayer has survived floods and fires; the boom times and the lean times; the coming of the railroad and the departure of the railroad; the coming of the super highway and the abandonment of Main Street; and the influx of people when times were good and the decrease in the population when they were not. But Mayer has persevered, just as the miners, ranchers, merchants and residents of Mayer did in Joe Mayer's day. In an article about Mayer by editor Dwight L. Bockman in the August 12, 1966, issue of the *Desert View*, Bockman wrote:

> This is my town. I can wave at my neighbor, yell at my dog, sleep with my doors unlocked, kibitz with the post office "social hour" or listen to the tall tales at Orin Culver's service station.
>
> I go out to pick up a one-inch ad, stop and visit a couple of hours, take a run out to a rock quarry, even bring some rocks back with me. Oh, the ad! Almost drove off without it.
>
> Stopped in Earl Vaughn's store for a pipe fitting, heard an interesting "new age" conversation going on, so sat down and made myself at home for awhile.
>
> Turned my car around to go back home, saw a couple of chamber of commerce members, stopped and helped them put a wheel on a wagon they were fixing up for a yard piece. Ended up with them in Duncan's Oak Hill Restaurant for coffee, over which we indulged in some big talk about the possibilities of a museum here in Mayer, as well as other things for the welfare of the community.
>
> When am I going to get my work done? Well, I'll work on the paper tonight. I've been on the night shift so many years, guess I do my best work at night.

All in all, Mayer has not changed a great deal since Joe Mayer found the place he would call home in 1882. As Winnie Mayer Thorpe said, "Mayer hasn't changed all that much, maybe just the clothes." Joe Mayer would be welcome to sit on the porch of the Mayer hotel and watch the daily life of his town go by. Photograph by Nancy Burgess, 2011.

Then I come out of my shop after midnight. I look up at the beautiful star-studded sky, fill my lungs with krisp [sic] fresh air, and I know this is my town.
What is this town to me? It is contentment. It is the compromise of the extremes. Let it ever go forward, to prosper and develop into something finer, and finer, and finer. My town.

Mayer today rests on its laurels in many ways as "Joe Mayer's Town." But Joe Mayer died over a century ago, and in those 100-plus years, the town has evolved into a modern community with a much-appreciated historic past. In spite of the ups and downs of any small town, it has gotten, as Dwight Bockman wrote, "finer, and finer, and finer." If Joe Mayer were to walk down Central Avenue today, he would soon know right where he was—right in the center of Joe Mayer's town. He could step right up onto the porch of the Mayer Hotel, have a seat, put his feet up on the railing and watch the daily life of "Joe Mayer's Town" go by.

Bibliography

Archives
Arizona Department of Library and Archives, Phoenix, AZ. Various records.
Arizona Historical Foundation, Arizona State University, Tempe, AZ. Library and archives collections.
Arizona Historical Society, Tucson, AZ. Library and archives collections.
Arizona State University, Hayden Library, Tempe, AZ. Archives special collections.
Sharlot Hall Museum, Prescott, AZ. Library and archives collections.
State of Arizona, Phoenix, AZ. Library and archives collections.
Yavapai County Assessor's and Recorder's Offices, Prescott, AZ. Various records.

Newspapers
Arizona Farmer (Phoenix, AZ). Supplement of the *Arizona Gazette*. Various dates.
Arizona Gazette (Phoenix, AZ). February 14 and April 17, 1905.
Arizona Miner (Prescott, AZ). Various dates.
Arizona Republic (Phoenix, AZ). Various dates.
Arizona Republican (Phoenix, AZ). Various dates.
Arizona Weekly Journal-Miner (Prescott, AZ). Various dates.
Arizona Weekly Miner (Prescott, AZ). March 19, 1900.
Big Bug Breeze (Mayer, AZ). Various dates.
Big Bug Copper News (Mayer, AZ). December 1917.
Bisbee Daily Review (Bisbee, AZ). August 2, 1907.
Coconino Sun (Flagstaff, AZ). Various dates.
Courier (Prescott, AZ). Various dates.
Daily Courier (Prescott, AZ). Various dates.
Daily Journal-Miner (Prescott, AZ). September 11, 1902.
Desert View (Mayer, AZ). Various dates.
Florence Blade Tribune (Florence, AZ). November 1911.
Journal Miner (Prescott, AZ). Various dates.
Mayer Miner (Mayer, AZ). Various dates.
Phoenix Gazette (Phoenix, AZ). Various dates.
Prescott Daily Courier (Prescott, AZ). Various dates.
Prescott Evening Courier (Prescott, AZ). Various dates.
Prescott Sun (Prescott, AZ). April 7, 1993.
Prospect (Prescott, AZ). Various dates.
Rio Abajo Weekly Press (Santa Fe, NM). March 31, 1863.
Salt River Herald (Phoenix, AZ) March 2, 1878.

See Scenic Southwest, vol. 11, no. 6, June 1939.
The Paper (Prescott, AZ). July 29, 1976.
Yavapai (Prescott, AZ). Various dates.
Yavapai County Messenger (Prescott, AZ). Various dates.

Publications

Arizona Directory Company. *Yavapai County Directory*. Phoenix, AZ: 1916.
Arizona Good Roads Association Illustrated Road Maps and Tour Book, 1913. Phoenix: Arizona Highways, Arizona Department of Transportation, Reprinted 1987.
Arizona Highways. "Arizona's Impossible Railroad." Phoenix: Arizona Department of Transportation, September 1979.
Arizona State Business Directory. Phoenix, AZ: 1912, 1915–1916, 1916–1917.
Barnes, Will C. "The Black Canyon Stage," *Arizona Historical Review*, vol. 6, no. 2, April 1935.
Bartz, James Lynn. *Company Property of Wells, Fargo & Co's. Express 1852–1912*. Lake Forest, CA: The Westbound Stage, 1993.
Bechtel, Robert B., and Mynne Cordes Jarman. "Cordes and Cordes Junction," *Journal of Arizona History*, vol. 26, no. 4, Winter 1985.
Bensch, Dr. Ernest. *Untitled Article*. Unpublished: N.D.
Boyer, Jay. *Coconino County Center for the Arts Presents Western Films, The Big Screen and the Big Sky (chapbook)*. Flagstaff, AZ: Coconino Center for the Arts, N.D.
Burton, Jeffrey F., Mary M. Farrell, Florence B. Lord and Richard W. Lord. *Confinement and Ethnicity: An Overview of World War II Japanese American Relocation Sites*. Washington, D.C.: Western Archaeological and Conservation Center, National Park Service, U.S. Department of the Interior, 1999.
Canty, J. Michael, and Michael N. Greeley, Editors. *History of Mining in Arizona, Volumes I, II*. Tucson, AZ: Mining Club of the Southwest Foundation and American Institute of Mining Engineers, 1987, 1991.
Collins, Willliam S. *The New Deal in Arizona*. Phoenix: Arizona State Parks Board, 1999.
Cordes, Claire Champie. *Ranch Trails and Short Tales*. Crown King, AZ: Crown King Press, 1991.
Culley, Matt. "Old Yavapai is Cattle Country." *Arizona Highways*, April, 1964.
Dedera, Don. Letter to Bud Melcher, February 25, 1969.
Dow, James M., and Alfred "Bud" Francis to Don Dedera. Bill of Sale, October 15, 1962.
Eskes, Dave. "Feminine Frontier." *Arizona Highways*, April 2007.
Fagerberg, Jr., Dixon. *Meeting the Four O'Clock Train and Other Stories*. Prescott, AZ: Sharlot Hall Museum Press, 1983.
Farish, Thomas Edwin. *History of Arizona Volume I*. Phoenix: Printed and Published by direction of the Second Legislature of the State of Arizona, 1915.
Fritz, Scott. "Yavapai County Merchants: the Center of Arizona's Early Territorial Economy, 1863–1881," *Territorial Times*, vol. 1, no. 2, May 2008.
Granger, Byrd H. *Will C. Barnes' Arizona Place Names*. Tucson: University of Arizona Press, 1960.
Hall, Sharlot M. *Cactus and Pine*. Second Edition. Phoenix, AZ: Sharlot M. Hall, 1924.
_____. *Poems of a Ranch Woman*. Second Edition. Prescott, AZ: Sharlot Hall Historical Society of Arizona, 1989.
Hanchett, Leland L., Jr. *Catch the Stage to Phoenix*. Phoenix, AZ: Leland L. Hanchett, Jr., 1998.

Haskett, Bert. "History of the Sheep Industry In Arizona," *Arizona Historical Review*, vol. 7, no. 3, July 1936.
Hinton, Richard J. *Handbook to Arizona 1877*. Second Edition. Glorieta, NM: Rio Grande Press, Inc., 1970.
Kirkman, Marshall M. *Building and Repairing Railways, Supplement to The Science of Railways*. New York: The World Railway Publishing Company, 1902.
Lee, Antoinette J., ed. *CRM: The Journal of Heritage Stewardship*. Washington, D.C.: The National Park Service, vol. I, no. 2, Summer 2004.
Lindgren, Waldemar. *Ore Deposits of Jerome and the Bradshaw Mountains Quadrangles, Arizona*. GPO USGS Bulletin 782, Washington, D.C.: United States Geologic Survey, 1926.
London, Jack. *Call of the Wild*. New York: Grosset and Dunlap, 1903.
Luchetti, Cathy, and Carol Olwell. *Women of the West*. New York: The Library of the American West, Herman J. Viola, ed, Orion Books, 1982.
Lynch, Richard. *A Short History of Railroading in the Bradshaws, the Era of the Prescott and Eastern*. Unpublished manuscript, 1972.
McCraine, Kathy. *Cow Country Cooking*. Prescott, AZ: Kathy McCraine, 2010.
McCroskey, Mona Lange. "The Most Efficient Fiber Producers on Earth: Angora Goat Ranching in Yavapai County, Arizona, 1880–1945," *Territorial Times*, vol. 2, no. 1, November, 2008.
Melton, Brad, and Dean Smith, eds. *Arizona Goes to War*. Tucson: University of Arizona Press, 2003.
Morgan, Ann Hodges, and Rennard Strickland. *Arizona Memories*. Tucson: The University of Arizona Press, 1984.
Morgan, Lerah Cooper, Editor. *Echoes of the Past Volume 1*. Prescott: The Yavapai Cowbelles of Arizona, 1955.
Myrick, David F. *Santa Fe to Phoenix, Railroads of Arizona, The Santa Fe Route, Volume 4*. Berkeley, CA: Signature Press, 1998.
_____. *Santa Fe to Phoenix, Railroads of Arizona, The Santa Fe Route, Volume 5*. Berkeley, CA: Signature Press, 2001.
Phillips, Melvin W. *Mile High Docs*. Prescott, AZ: M & J Publishing Company, 1996.
Potter, Alvina M. *Many Lives of the Lynx*. Prescott, AZ: Alvina M. Potter, 1964.
Prescott City Directory. Long Beach, CA: The Inskeep Company, various dates.
Robinson, William Henry. *The Story of Arizona*. Phoenix, AZ: The Berryhill Company, 1919.
Rosebrook, Jeb J., and Jeb S. Rosebrook. "The Old Black Canyon Highway," *Arizona Highways*, August 1994.
Rozum, Fred A. "Buckboards and Stagecoaches: Establishing Public Transportation on the Black Canyon Route," *Journal of Arizona History*, vol. 30, no. 2, Summer 1989.
Sacks, Benjamin. *Be It Enacted: The Creation of the Territory of Arizona*. Phoenix: Arizona Historical Foundation, 1964.
Sagstetter, William E., and Elizabeth M Sagstetter. *The Mining Camps Speak*. Denver: BenchMark Publishing of Colorado, 1998.
Samuelson, Susan Adams. "The Orme School on the Quarter Circle V Bar Ranch," *Journal of Arizona History*, vol. 25, no. 4, Winter 1984.
Santa Fe, The Chief Way Reference Series, "System Standards, Volume Two." Dallas: Kachina Press, 1978.
Sayre, John W. *Ghost Railroads of Central Arizona*. Phoenix, AZ: Red Rock Publishing Company, 1985.
_____. *The Santa Fe, Prescott & Phoenix Railway, the Scenic Line of Arizona*. Boulder, CO: Pruett Publishing Company, 1990.

Schweikart, Larry. *A History of Banking in Arizona*. Tucson: The University of Arizona Press, 1982.

Sheridan, Thomas E. *Arizona A History*. Tucson: The University of Arizona Press, 1995.

Smith, Dean, ed. *Arizona Highways Album, The Road to Statehood*. Phoenix: Arizona Department of Transportation, 1987.

Sparkes, Grace M. *Yavapai*. "The Land of Opportunity." *University of Arizona Bulletin*. Bulletin no. 59, County Resource Series No. 2, 1917–18.

Spude, Robert L. "A Shoestring Railroad: The Prescott and Arizona Central," *Arizona and the West*, vol. 17, no. 3, Autumn 1975.

_____. "A Land of Sunshine and Silver," *Journal of Arizona History*, vol. 16, no. 1, Spring, 1975.

_____. *The Poland Branch of the Bradshaw Mountain Railroad*. Unpublished manuscript. Tempe: Arizona Historical Foundation, N. D.

Stevens, Horace J., Compiler. *The Copper Handbook, Volume VI*. Houghton, MI: Horace J. Stevens, 1906.

_____. *The Copper Handbook, Volume VII*. Houghton, MI: Horace J. Stevens, 1907.

Stevens, Robert C., Editor. *Echoes of the Past, Tales of Old Yavapai, Volume 2*. Prescott, AZ: The Yavapai Cowbelles, 1964.

Stein, Pat H., and Elizabeth J. Skinner. *Mining the Big Bug: Archaeological Investigations at Twelve Historic Sites between Mayer and Dewey, Yavapai County, Arizona*. Flagstaff, AZ: SWCA, Inc., 1997.

Theobald, John, and Lillian Theobald. *Arizona Territory Post Offices and Postmasters*. Phoenix: The Arizona Historical Foundation, 1961.

_____. *Wells Fargo In Arizona Territory*. Phoenix: The Arizona Historical Foundation, 1978.

Thorpe, Winnifred L. Mayer, notes by Robert L. Spude. "Joe Mayer and His Town," *The Journal of Arizona History*, vol.19, no. 2, Summer 1978.

Trimble, Marshall. *Diamond in the Rough, An Illustrated History of Arizona*. Norfolk, VA: The Donning Company, 1988.

United States Census. Yavapai County, AZ, 1880, 1900, 1910, 1920, 1930.

United States Civilian Conservation Corp. *The Official Annual for 1936, Phoenix District, 8th Corp Area*. Baton Rouge, LA: Direct Advertising Company, 1936.

Walker, Henry P., and Don Bufkin. *Historical Atlas of Arizona, Second Edition*. Norman: University of Oklahoma Press, 1986.

Wagoner, Jay J. *Arizona Territory 1863–1912, a Political History*. Tucson: The University of Arizona Press, 1980.

Westerlund, John S. *Arizona's War Town*. Tucson: The University of Arizona Press, 2003.

Wilson, Eldred D. *Gold Placers and Placering in Arizona*. Tucson: Arizona Bureau of Geology and Mineral Technology, Geological Survey Branch, Bulletin No. 168, 1961.

_____, J. B. Cunningham and G. M. Butler. *Arizona Lode Gold Mines and Gold Mining*. Tucson: Arizona Bureau of Geology and Mineral Technology, Geological Survey Branch, Bulletin No. 137, 1934 (revised 1967).

Willson, Roscoe G. *Pioneer and Well Known Cattlemen of Arizona*. Phoenix, AZ: Valley National Bank, 1956.

Wright, Nancy Kirkpatrick. *Sharlot Herself*. Prescott, AZ: Sharlot Hall Museum, 1992.

Yavapai Commercial Club. *Yavapai County Arizona: The Treasure Vault of the Southwest*. Chicago: Poole Bros., 1907.

Zanjani, Sally. *The Glory Days in Goldfield, Nevada*. Reno: The University of Nevada Press, 2002.

Index

Abaja 9; *see also* Fort McDowell Yavapai Nation
Addicks, J.E. 136–139, 155, 172, 175, 177, 180
Adney, Mrs. Julia 130
Agua Fria River 1, 3 17, 20, 35, 58, 62, 66, 247, 249
Agua Fria Station (Spaulding Station) (Dewey) 20
Agua Fria Valley 35, 42, 53, 174
Alabam Freight Lines 122
American Laundry 46
American Onyx Products, Inc. 150
Annie Mine 156
Antelope Creek 58, 142
Antelope Station *see* Cordes
Apache People 9, 10
Arcosanti 251
Arctic Ice and Meat Company 51, 118, 124
Arizona (state) 9–13, 24, 29, 30, 35, 42–43, 51, 53–55, 57, 59, 65, 71, 73, 78, 04, 110, 118–120, 127, 132, 135–136, 138–139, 141–142, 155, 166, 168, 171, 173, 182, 186–187, 193–196, 198–200, 202, 205, 209–212, 214–215, 217, 219–221, 231, 238, 249
Arizona City (Curtiss) 156, 158–160, 174, 245, 247
Arizona Cooler 117–118
Arizona Goes to War 211, 214–215
Arizona Highway Department 23–24
Arizona Highways 65–66, 73
Arizona Highways Album: The Road to Statehood 195
Arizona Miner 104, 145
Arizona Mining Supply 144
Arizona National Guard 213
Arizona Onyx Company 182
Arizona Power Company 51, 154
Arizona Republic 166, 218, 221, 227

Arizona Republican 29, 33, 53, 77, 96–99, 146, 174, 195
Arizona State University 81
Arizona Territory (Territory of Arizona) 11–12, 16, 27, 61, 75, 78, 110, 142, 169, 171, 181, 193–194, 196
Arizona Territory Post Offices and Postmasters 6
Arizona Weekly-Journal Miner 139, 156
Arizona Weekly Miner 172
Arizona's War Town 212–213
Ash Fork 3, 35, 83, 171–172, 174
Atchison, Topeka & Santa Fe Railway 171, 175, 182, 188
Atlantic & Pacific Railroad 27, 35, 171
Auther, William 192

Bagdad 166
Baker, James 63
Baldwin, Dr. Warren 130
Basketmakers, Yavapai Indians 10, 51
Batre, Charles 126
Bechtel, Robert B.: *Cordes and Cordes Junction* 69
Benedict, Principal Joel A. 84–86
Bensch, Dr. Ernest 66
Bensch Ranch 66, 249
Big Bug (town) 21, 155, 160
Big Bug Breeze 102, 116, 118, 122
Big Bug Copper News 126, 130
Big Bug Creek 2, 20–21, 28–29, 35, 42, 44, 50, 58, 74, 76, 93, 94, 119, 143, 144, 153, 155–156, 158–159, 163, 174, 179, 181, 183–184, 186, 198, 206, 212, 222, 224, 238,-239, 245, 247
Big Bug Creek Flood, February, 1891 94
Big Bug Mine 40, 139, 142
Big Bug Mining and Milling Company 141–142

Big Bug Mining District 10–11, 35, 43–44, 59, 99, 143, 144, 154, 172, 175, 179
Big Bug Reduction and Development Company 159
Big Bug School, Mayer 76–77, 79
Big Bug Smelter (Boggs Smelter) 159–160
Big Bug Station (Stage Stop) 1, 2, 5, 20–21, 25, 27–29, 39, 77, 92, 113, 147, 245
Big Ledge Mining Company 162–164, 183
Bill Thompson Ranch 67
Bittner, Elladean 65
Black Canyon City (Canon) 1, 16, 227, 244, 249
Black Canyon Highway 16, 23–24, 60, 66, 190, 212–213, 216, 218–219, 221–222, 226, 239, 244, 248
Black Canyon Road 3, 9, 14–19, 21–24, 27, 29, 35–36, 39, 41, 96, 99, 121, 143, 163, 174, 177, 183, 206, 225, 234–235, 244, 249
Black Canyon Route 14, 168
Black Canyon Stage Line 1, 5, 16
Block, Ed 152
Bockman, Dwight: *Desert View* 227–228, 251, 253
Boggs, Theodore Warner 155, 156, 162
Boggs Mine/Smelter (Big Bug Smelter) 148, 156, 159, 245, 247
Bolding's Market 231
Born, Dewey 234
Boundary, Arizona and New Mexico 12
Boundary, U.S. and Mexico 12
Bowers' Ranch (Edward) 42, 62, 174
Boyer, Jay: *The Big Screen and the Big Sky (Chapbook)* 231–232, 234

259

Index

Bradshaw Mountain Railroad 2, 45, 152, 179, 183, 199, 217
Bradshaw Mountain Railroad, Crown King Branch 44, 164, 179–182, 248–248, 251
Bradshaw Mountain Railroad, Poland Branch 143, 153, 181
Bradshaw Mountains 1, 9–11, 16, 27, 35, 39, 42, 44, 58, 65, 131, 141, 165, 184, 202, 208, 221, 238, 251
Breezy Pines 182, 249
Brickyard, Mayer 35, 103–104
Brooks Locomotive Company 167, 171
Brown, Angie 58, 65
Brown, George E. 58, 65
Bumble Bee 1, 3, 20, 44, 175, 220, 249
Butternut Mine 143, 154, 156

Calderwood, Capt. M.H. 20
Calderwood Butte or Peak 20
California 12, 14, 28, 36, 38, 142
California, Arizona & Santa Fe Railway Company 188
Call of the Wild 45
Cameron, Rittie McNary 65
Camp Verde 9, 14, 221, 224
Campbell and Francis Sheep Company 70
Caniglia, Noel 74
Caniglia, Tommy 74
Carleton, Brig. Gen. James H. 13
Carlson, Raymond: *The Three Rs on the Range* 90
Catch the Stage to Phoenix 17
Cather, Willa: *Song of the Lark* 5
Catholic Church, Mayer 131–132, 224–245
Cedar Canyon 67
Charles H. Hooker Ranch 89
Cherry Creek (Dewey) 35, 42, 174, 176–177,
Chinese in Mayer 29, 47, 49, 94, 113, 116, 124, 133
Civilian Conservation Corps (CCC) 201–204, 208–209, 211, 224
Cleator (Turkey Creek) 44, 133, 227, 249
Coconino Sun 160
Collins, Mrs. J.M. 83
Collins, William S.: *The New Deal in Arizona* 200, 202
Colorado 25–26
Colorado River 12
Commercial Mining Company 156, 159–161
Concord Stage 17
Connors, Mike 241

Cook, Dean 2
Cook, Joe 191
Cook, Rhonda 235–237
Cook, Rob 235–237
Copper Handbook Vol. VI 162, 164
Copper Mountain 94
Cordes (Antelope Station) 1, 3, 20, 66, 69–71, 73, 84, 249
Cordes, Clair Champie 65, 133
Cordes Junction 69, 221–222, 227, 251
Cottonwood 244
Cottonwood Station 21
Coughan, Mrs. 231
Courier 117, 151, 245, 247
Cow Country Cooking 74
Crawford, Noble Ray "Ray" 220, 223
Crawford, Tom 220, 223
Crazy Basin 1, 3, 44, 138
Crook, Gen. George S. 9
Crook Canyon 209
Cross U Ranch 74
Crown King 2, 19, 67, 97–98, 127, 133, 141, 175, 180–181, 248–249, 251
Crowned King Mine 98, 141, 179
Culley, Matt: *Old Yavapai is Cattle Country* 64, 66–67
Culver, Orin 231
Curtiss (Arizona City) 156, 158, 174

Daily Journal Miner 161, 186, 195
Davis, Ish 119, 132
Davis, Jesse 149
Dedera, Don 190–191, 195; *Arizona Highways Album: The Road to Statehood* 195
Deeds, W.S. 103
de Espejo, Don Antonio 11
de los Godos, Capt. Marcos Farfan 11
de Onate, Juan 11
Desert View 227–229, 231, 251
De Soto Mine 44
Dewey (Spaulding Station) (Agua Fria Station) (Cherry) 16, 20, 57, 62, 148, 174, 176–177, 219, 247–249
Dewey-Humboldt 57, 60, 221, 237, 244, 249
Diamond in the Rough, An Illustrated History of Arizona 91
Dilzhe'e 9; *see also* Tonto Apache
Diskin, Pete 98
Douglas 194
Douglas, Prof. James 35, 104, 159

Douglas, James Stuart "Rawhide Jimmy" 35, 104
Drake, William A. 99, 172
Dugas, Fred 52, 66, 121
Dugas Accommodation School 89
Dugas Ranch 52, 66, 89
Duncan's Oak Hills Motel/Restaurant 227–228, 230–231, 243, 251
Dust in Our Desks 78–79

Echoes of the Past, Tales of Old Yavapai, Vol. 1 54, 110
Elder, James B. 113
Elliott, W.S. 83
Entro 174; *see also* Prescott Junction (P&E Junction)
Eskes, Dave: *Arizona Highways* 65
Eugenie 152
Eugenie Mine 153

Fain & Heath 65
Fain Family 57, 66
Farfan de Los Godos, Marcos 11
Farish, Thomas Edwin: *The History of Arizona Vol. 1* 136
Fireman, Bert 54
First Schoolhouse, Prescott 75, 77
Fisher, E.A. 126
Flagstaff 2, 3, 31, 71, 214–215, 231
Flinn, Dr. John 46
Florence Blade-Tribune 194
Ford Times: The Three Rs on the Range 90
Fort McDowell 1
Fort McDowell Yavapai Nation 9
Fort Verde 1
Fort Whipple 1, 13, 27
Frazer, W.A. 52, 113
Fred Venator's Market/ Slaughter House 113
Fremont, Jessie Benton 81
Frémont, Gov. John C. 81–82
Fritz, Scott: *Yavapai County Merchants: The Center of Arizona's Early Economy, 1863–1881* 109

G & S Gravel, Inc. 245–246
Gadsden Purchase 12, 136, 195
Gage, E.B. 187
Garber, J.E. 86
Gardner, Gail 110
Gardner, James I. 110
Garrett, Frank J. 123, 124, 205, 213
George, William "Bill" 111
George A. Treadwell Mining

Index

Company *see* Treadwell Mining Company
Ghost Railroads of Central Arizona 35, 42, 161, 179, 187
Gila County 215
Gila River 12, 199
Gillette (Gillett) 1, 17, 18, 20
Gillette, D.B., Jr. 17, 18
Givens, Jack 151
Givens Stone Yard 231
Gleeson 41
Globe (Globe City) 26–27
Globe Mine 26
Gold Placers and Placering in Arizona 145
Goldthwaite, W.S. 52
Goldwater, Barry M. 6
Goodwin, Gov. John N. 75
Goswicks 66
Governor's Mansion, Prescott 58
Granite Creek 77, 142
Granite Dells (Point of Rocks) 22, 34–35, 42, 168, 174–175, 232
Grant, Pres. Ulysses S. 78
Grant Bros. 48, 113
Grapevine Canyon/Springs 104, 160, 224
Grapevine Creek 35, 45
Great Depression 63, 89, 199–201, 206, 212
Great Western Smelter Company 148, 162–164, 183
Great Western Smelter Smokestack *see* Mayer Smokestack
Greyhound Bus, Mayer 216

Hackberry & Boggs Railroad (Hackberry Railroad) 156, 158–161
Hackberry & Iron Queen Railroad 160
Hackberry Mining Company/Hackberry Mine 148, 156, 159, 160,
Hall, Adelaide 57, 63
Hall, Dr. E.A. 129
Hall, James 57, 63
Hall, Sharlot Mabridth 7, 54–57, 63, 195; *Poems of a Ranch Woman* 57
Hanchett, Leland L., Jr.: *Catch the Stage to Phoenix* 17
Harris, Postmaster J.E. 131, 207
Harts of the West 233, 235
Hassayampa River 94
Hassayampers 53, 136, 138, 141,
Heffleman, Henry C. 60, 113, 119
Hegglin, Martha 116–117
Henrietta Mine 152, 162

Henrietta Station (Torres) 152
Hickey, Martha Hegglin 110, 117–118
Highway (High Way) Garage 122–124, 205, 212–213, 219, 226
Highwaymen 16
Hill, James E. 53
History of Arizona Vol. 1 136
Hitching Post Market 229
Hoffmire, Antoine 25
Hoffmire (Hoffmayer, Hoffmeier, Hoffmeyer), Joseph 25
Hooker, Charles 89
Hooker Accommodation School 89
Hooker and Kellogg 65
Hopkins, Ernest J.: *Dust in Our Desks* 78
Horrmire, Marie Therese 25
Hospitals 38, 48–49, 113, 127–130
Humboldt (Val Verde) 15, 42, 56, 59, 69, 112, 127, 143, 144, 148, 161, 174, 184, 187, 195, 219, 227, 249
Humbug Creek 94
Hunt, Gov. George W.P. 38, 49, 194–195
Hunting 9, 59, 229, 231
Huron 35, 42, 143, 148, 152, 174, 179, 186–187
Huron Mines 179, 187
hydraulic mining 92–94

Indian Census, 1910 50–51
Indian Souvenir Toothpick Company 106
International Onyx and Marble Company 150, 182
Iron King Mine 146
Iron Queen Mine 148, 160
Ish Davis' Saloon 118–119

Jaeger, John A. 180
Jarman, Mynne Cordes: *Cordes and Cordes Junction* 69
Joe Mayer Club 118
Johnston, Georgiana 182
Johnston, Laura 231
Johnston, Mildred 182
Jordan Stages 120
Joseph Cook's Saloon 118–119
Journal Miner 19, 21, 29, 38, 42–43, 47, 67–68, 83, 94, 96–97, 101, 105–107, 129–130, 139, 149–150, 155–156, 161, 167, 173, 179, 184, 186–187, 195, 197
Judd, Bert A.: *The Man Behind the Pick* 135

Kelly, Roger E.: *"America's*

World War II Homefront Heritage" 211
Knapp, Martin T. 38, 53, 97
Koogler, Bernice Jane 216
Kung, Charley (or Fong, Charley) 49

Ladies' Aid Building 81–83, 86, 87, 165, 200, 251
LaDue, Ernest 126
Laura's Café 231
Lee, Sam 46, 124
Lessard, Clare 131
Lessard Ranch 56
Lev Nellis' Meat Market 36, 124, 188
libraries 131, 133, 239, 244
Lincoln, Pres. Abraham 12
Little Jessie Mine/Mill 155–156
Livingstone, John 99
Lloyd, Dr. 1
La Loma del Cobre 84, 113
London, Jack: *Call of the Wild* 145
Lonesome Valley 35, 42, 56–58, 60, 62, 66, 174, 177
Long, E.I. 83–84
Long Beach, California 12, 36, 38
Looney, Marjorie Belle 36, 38
Looney, Dr. Robert Nelson 36, 38, 127–128
Lynx Creek 12, 35, 42, 50, 62, 92–94, 142–143, 152, 154, 174
Lynx Creek Mining District 181
Lynx Creek Reservoir 92

Mahoney, Ralph: *Motorists Sigh with Relief as New Road Nears Reality....* 218–219, 227
The Man Behind the Pick 135
Many Lives of the Lynx 144
Maps 143, 148, 157, 170, 178, 183, 203, 221
Maricopa County 9
Marion, John 14
Marr's Ranch (John L.) 62
Martin, John A. 130
Maxwell, Margaret: *The Depression Years in Yavapai County* 199–200
Mayer, Joseph "Joe" 2, 5–8, 20, 21, 25, 27–29, 32, 33, 35–36, 38–39, 42–47, 50–51, 53, 60, 76–77, 81–83, 91, 93–94, 96–99, 101–109, 113, 118, 120–121, 127, 146–149, 153, 155, 160–162, 166, 172, 179, 182, 186–187, 224, 226–227, 231, 242, 253

Mayer, Joseph H. 38, 115
Mayer, Martha Gertrude "Martie" 5, 27–28, 30, 36, 38, 76–77, 127–128, 147
Mayer, Mary Belle "Mamie" 5, 26–27, 30, 36, 38, 49, 76–77, 124–125, 130–131, 147, 206, 207, 226
Mayer, Sarah Annie Skelton "Annie" 38, 49, 115, 125, 244
Mayer, Sarah Belle Wilbur "Sadie" 2, 5–8, 17, 25–27, 29, 31–33, 38–39, 49, 51, 60, 93–94, 105–106, 115, 125–127, 130–132, 147, 206, 207, 230, 251
Mayer, Wilbur Joseph "Burr" 5, 31–32, 38, 49, 76–77, 103, 113, 115, 119, 120, 122, 125, 147, 244
Mayer, Wilbur Nelson "Nelson" 38, 113, 125
Mayer, Winifred Lucille "Winnie" 5, 25–26, 29, 31–32, 37–38, 50–51, 60, 76, 94, 99, 101–102, 104, 106, 113, 115, 120, 130–131, 133–134, 147, 252
Mayer & Mayer 11, 49, 53, 104, 113, 119
Mayer Apartments 7, 35, 103–105, 124, 230–231, 245
Mayer Barber Shop 35, 104
Mayer Blacksmith Shop 47
Mayer Business Block 7, 11, 34–37, 41, 43, 91, 100, 103–104, 109, 11, 113, 125, 130–131, 147, 183, 197–198, 207, 215–216, 218, 220, 225–226, 230–231, 233–236, 240–241, 245
Mayer CCC Camp F-33A 202–204
Mayer Chamber of Commerce 230
Mayer Custom Plant 164
Mayer Daze 234–237
Mayer Depot 8, 50, 68, 111, 165, 179, 176, 185, 189–192, 198, 213
Mayer Elder's Club 231, 237
Mayer Evacuation Center 210
Mayer Freighting Company 99
Mayer Garage 120–121
Mayer High School, Mayer 86, 113, 206, 207, 224
Mayer High School, Spring Valley 86, 250–251
Mayer Hospital 48–49, 113, 129–130
Mayer Hotel 11, 21, 34–35, 41–45, 47–48, 53, 60, 94, 96–97, 102–104, 111, 113, 119, 121–122, 124, 127, 177, 186, 198, 207–208, 212–213, 218–219, 235–236, 240, 242, 245, 252–253
Mayer Library 131, 133, 239, 244
Mayer Livery, Feed and Sales Stables 111
Mayer Mercantile 35, 91, 99, 104, 113, 115, 205
Mayer Miner 53, 119
Mayer Mining and Milling Company 164
Mayer Onyx Mine/Quarry 98, 149–150; *see also* Onyx Quarry
Mayer Ore Purchasing Plant/Company 149, 163–164
Mayer Owl 34, 36, 43, 104, 120–121, 123–125, 131, 188, 204, 206, 212
Mayer Post Offices 32–33, 49, 104, 130–131, 183, 212, 218, 225, 231, 235, 240, 251
Mayer Saloon 35, 91, 103, 104, 109
Mayer School (1902) 30, 31, 76–77, 87, 250; *see also* Ladies' Aid Building
Mayer School (Modern) 251
Mayer School District 77, 82–84, 86, 191
Mayer School Literary Society 84
Mayer Smelters Corporation 163
Mayer Smokestack 2, 164, 247
Mayer State Bank 103–104, 125–127, 199, 212, 240, 245
Mayer Station (Mayer's) 2, 20–21, 28–29, 32, 37, 43, 81–82, 91, 94–95, 103–104, 111, 113, 126, 131, 147, 177, 207, 212, 217, 222, 229, 231, 251
Mayer Water Company 53, 199, 224
Mayer's *see* Mayer Station
McCabe 38, 127, 144, 155–156, 187
McCann, Al 149
McCraine, Kathy: *Cow Country Cooking* 74
McCroskey, Mona Lange: *The Most Efficient Fiber Producers on Earth: Angora Goat Ranching in Yavapai County, Arizona, 1880–1945* 73
McDonald, Clyde 118, 241
McDonald, Irene Thompson 83, 131, 227
McDonald, Sonya 118, 241
McFarland, Sen. Ernest W. 215
McMichael, J.H.M. "Harry" 120, 124

McMillan Camp 27
Meany, Mabel 77
Mercy Hospital, Prescott 127–128
Mexican War of 1848 12
Mexico 12, 71, 136
Mining the Big Bug: Archaeological Investigations at Twelve Historic Sites Between Mayer and Dewey, Yavapai County, Arizona 11
Minnesota and Arizona Construction Company 172
Minucci, Archie 123–124
Minucci, Rosena Promberger 124–125, 206
Mr. Yoeman's Wood Yard 50
Mix, Tom 232, 234
Mogollon Rim 71
Mohair 66, 72–73
Monihon, James D. 17
Moore, Hill C. 42, 94
Moore, Nellie 65
Mountain Men 11, 12
Mueller, Emil 131, 223, 228, 231
Muncey (Munsey or Muncy), William M. 2, 21, 21–28, 227
Munds, Sheriff John "Johnny" 30–31
Murphy, Frank M. 2, 35, 42, 44, 99, 149, 151–152, 154–155, 169, 171–172, 175, 179, 181, 187
Murphy, Gov. Nathan Oakes 169, 193
Myrick, David: *Santa Fe to Phoenix, Railroads of Arizona Vol. 5* 156, 159, 161, 171, 179, 183

National Register of Historic Places 104–105, 230–231, 242
Nellis, Leverette P. "Lev" 36, 124, 131, 188
Nelson Tavern and Amusement Hall 113
Nevada 12, 103, 104
New Deal in Arizona, The 200, 202
New Mexico (state) 12–13, 25–26, 61, 194–196
New Mexico Territory 61
New River 1–3, 16–17, 20
Northern Arizona Territorial Normal School 82
Norwood, Joe 98

Old Black Canyon Highway *see* Black Canyon Highway
Old Black Canyon Road *see* Black Canyon Road

O'Neill, William O. "Buckey" 31, 149, 182
Onyx 31–32, 96, 98, 105–106, 143, 144, 149–151, 179, 182–183, 224, 227, 231, 246–247
Orchard Ranch 54–58, 63, 66
Orme Ranch (Quarter Circle V Bar Brand) 64, 89–90, 251
Orme School 64, 66, 88, 89–90, 251
Oro Blanco Mines 44
Osborn, Gov. Sidney P. 215

Palatkwapi Trail 11
Patterson, Caldwell & Company 17
Peck Mine Road 21
Petit, Joe 83
Phelps-Dodge Company 35, 104
Phoenix 1–3, 5, 8–9, 14, 16–17, 20, 24, 30, 33, 35–36, 43, 53, 67–68, 89, 99, 102, 116, 146, 171–172, 174–176, 187, 189–192, 195, 203, 212, 214–215, 218–222, 228, 231, 236–237
Pierce, Principal Loren 84
Pine Grove Mining District 45
Pioneer History Museum 102, 104
Pioneer Placer Mining District 142
Plaza Stables 17, 19
Poems of a Ranch Woman 57
Point of Rocks *see* Granite Dells
Poland, Davis Robert 153
Poland Extension Gold Mining & Milling Company/Mine 153, 181
Poland Junction 84, 143, 148, 152, 154, 156, 159, 174, 181
Poland/Poland Depot 142–143, 144, 148, 151–152, 154, 181–183, 249
Polk, Frank 126–127
Porter, H.K. 159–160
Potter, Alvina N.: *Many Lives of the Lynx* 145
Prescott 1–3, 5, 9, 21, 142, 195–196, 198
Prescott & Eastern Railway (P&E) 2, 34–35, 42, 53, 60, 68, 101, 111, 120–121, 143, 147, 149, 152, 156, 160–161, 164–165, 167, 172–175, 177, 179–184, 186–189, 197–199, 213, 247
Prescott Courier 31, 83, 104, 120, 131, 133, 199, 227, 234
Prescott Electric Company 187
Prescott Free Academy 81–82
Prescott Junction (P&E Junction) (Entro) 34, 165, 168, 173–175
Prescott Sun 235
Price, Louis B. 122
Promberger, Caroline "Carrie" 125, 205
Promberger, Frank 123–124, 128
Promberger, Helen 125
Promberger, William "Bill" 85–86, 123–125, 205, 206, 207, 212, 218
Prospect 104, 135, 150, 152
Providence 152–153

Quarter Circle V Bar *see* Orme Ranch
Quesada, Alicia 65

Rafters 11 RV Park 57
ranching: angora goats 8, 73, 58, 68, 72; cattle 8, 40, 42, 45, 48, 61–63, 65–67, 72–74, 89–90, 179, 183, 197–199, 227; sheep 8, 42, 45–46, 48, 66–71, 73, 183, 199
Rawdin, J.E. 97
Red Brick Schoolhouse 82–86, 217, 251
Red Cross Day, Mayer 197
Red Rock Mine 153
Rice, Alfred E. 123–124
Rich, George 29
Rigby, Col. T. Johns 164
Rigby Mining and Reduction Works 53, 60, 104, 113, 119, 164–165, 183–184
Rio Abajo Weekly Press 13
Rio Grande 12
Roalsted, Felicite 216
Rock Springs 1–2, 24
Rogers, E.S. 105, 107
Rogers, Samuel C. "Charming Dale" 75, 77
Roosevelt, Pres. Theodore 201
Rough Rider(s)/Memorial Statue 105, 149, 196, 198

Safford, Gov. Anson P.K. 78, 80
saloons 78, 118–119
Salt River Herald 17
Salt River Valley 14, 71, 89, 199
Sam Lee Chinese Laundry 46, 124
Sam'l Hill Hardware Company 144
Samuelson, Susan Adams: *The Orme School and the Quarter Circle V Bar Ranch* 89
San Carlos Indian Reservation 9, 50
Sanitarium 129–130

Santa Fe, New Mexico 13
Santa Fe, Prescott & Phoenix Railway (SF, P&P) 27, 34–35, 42, 71, 99, 121, 167, 169, 171, 174, 179, 188, 192, 197
Santa Fe to Phoenix, Railroads of Arizona Vol. 5 156, 159, 161, 171, 179, 183
Santa Fe Trail 11, 39
Sayre, John W.: *Ghost Railroads of Central Arizona* 35, 42, 161, 179, 187
Scammel, George Byron 107, 161
Schrade, Louis Family 116
Scott, Ethel 231
Sears, Elden 221, 226
Selig Company 232, 234
Seligman 56
service stations 207, 223, 226, 228–229, 231, 251
Sharlot Herself 47, 57, 138
Sherman, Moses H. 80–82
Shrum, Dr. Riley 49
Silver City, New Mexico 25–26
Skinner, Elizabeth J.: *Mining the Big Bug: Archeological Investigations at Twelve Historic Sites Between Mayer and Dewey, Yavapai County* 11
Skull Valley 14, 171, 177
Sloan, Gov. Richard 194
snow 123, 220, 226
Solomon, Isador 1–2
Song of the Lark 5
Sparkes, Grace M.: *Yavapai, the Land of Opportunity* 136, 142, 200
Spaulding, Emma 20
Spauding, Ida 20
Spaulding Station (Agua Fria Station) (Dewey) 1, 20
stagecoach(es) 16–19, 22, 43–44, 111, 120, 146, 187, 235–236
stage(s) 20, 22, 120
Star Placer Claim 143, 160
State Highway 69 14, 23–24, 151, 164, 211, 217, 221, 223–224, 227–228, 230–231, 236–238, 240, 243–244, 247, 249, 251
State Highway 89 22, 23, 221
State Route 69 24
Stees, F.R. 21, 22
Stees (Steece), Fred 29
Stein, Pat H.: *Mining the Big Bug: Archeological Investigations at Twelve Historic Sites Between Mayer and Dewey, Yavapai County* 11
Stewart's T Anchor Ranch 65–66, 74, 111

stockyards, Mayer 45, 60, 66–67, 119, 183
Stoddard, Isaac Taft 94, 175
Stoneman, Col. George 14
Sunset Point 3
Surrett, Ellen 86
Swastika Mine 44, 117

Taft, Pres. William Howard 194–195
Taylor Grazing Act of 1934 71
telephone service, Mayer 48, 51, 97, 112–113
Tempe Normal School 81–82
Territorial Times 73, 109
Thomas, Alfred, Jr.: *Dust in Our Desks* 78
Thomas, D.W. 59
Thomas, "Uncle Dick" 53
Thorpe, Thomas Edward "Tom" 37–38
Thorpe, Thomas Edward, Jr. 38
Thorpe, Wilbur Robert 38
Thorpe, Winifred Lucille Mayer "Winnie" vii, 5–7, 25–26, 29, 31–32, 37–38, 50, 60, 76, 94, 99, 101–102, 104, 106, 113, 115, 120, 127, 130–131, 133–134, 252; *Joe Mayer and His Town* 5, 25, 38, 51, 94, 99, 101, 113, 127
Tiger Mine 44
Tip Top 5, 17, 26–28
Tip Top Mine 16, 18, 27
tokens 119, 124
Tonto Apache People 9
Torres (Henrietta Station) 152
Town of Dewey-Humboldt 57, 60, 221, 237, 244, 249
Town of Prescott Valley 35, 56, 221, 237, 244, 247
Treadwell, Edwin D. 97, 107, 160
Treadwell, Dr. George A. 44, 160–162
Treadwell Mining Company/Smelter/Mine 11, 44, 60, 108, 113, 160–164, 183
Treaty of Guadalupe Hidalgo 12
Trice, Charles E. 113
Trice & Morgan 113
Trimble, Marshall: *Diamond in the Rough, An Illustrated History of Arizona* 91

Turkey Creek (Cleator) 58, 133, 142, 147, 203

Underhill, George C. 149, 182
United States Census 21, 28, 46, 49, 113, 130, 238, 245, 247
United States Forest Service 73, 202, 249
United Verde Extension Mine 35, 104, 200–201
University of Arizona 38
Utah 12

Val Verde (Humboldt) 15, 184
Van Horn, Dr. James B. 130
Vaughn's 229, 231
Verde Valley 11, 14

Wagner, Dr. H.A. 164
Wagner (Wagoner) Hotel (White House Hotel) 48, 113, 115
Walker Mining District 153
Walker Party 142, 145
Walker-Poland Tunnel 153–154, 181–182
Walnut Grove Dam 28–29, 95
Walters, R.C. 125–126
Weaver, Pauline "Paulino" 12, 142
Weber Chimney Company 164
Weedin, Col. Thomas 194
Wells, Mary L. 53, 118–119, 133
Wells Fargo & Company Express 119, 187
Westerlund, John S.: *Arizona's War Town* 212
Western Union Telegraph Company 119, 187
Whipple Stage 22
White, Mrs. H.B. (Louise) 115
White House Hotel 2, 11, 45, 83, 115–117, 119, 147, 206, 218, 234, 240–241, 245
Wickenburg 1, 14, 16, 36, 171, 209
Wicks & Mayer 91, 97, 100–101, 104, 113
Wiener, Allison 192
Wilbur, Joshua 25
Wilbur, Martha Mary Young 25, 31–32, 60
Wilbur, Sarah Belle "Sadie" 25–26
Willard, Frances "Fanny" 30–31

William Fox Studio (Company) 232, 234
William Helling & Company 14
Williamson Valley 63
Wilson, Eldred: *Gold Placers and Placering in Arizona* 145
Wilson, Joseph W. 98
Wipukyipai (Yavapai) 9
Woolsey Ranch (King) 62, 174, 177
Workers of the World 199
Works Progress Administration (WPA) 86, 200
World War I 37–38, 73, 117, 119–120, 130, 155, 164, 183, 198–199, 228, 231
World War II 2, 63, 73, 86, 89, 193, 206, 211–217
Wright, Lonnie 102
Wright, Nancy Kirkpatrick: *Sharlot Herself* 55–57, 138
Wyoming 12

Yarnell Hill 3
Yavapai (Wipukyipai) 9, 10
Yavapai-Apache Nation 9
Yavapai Commercial Club 58, 60, 67, 150
Yavapai County 8–10, 12, 21–22, 24, 27, 30, 40, 46, 49, 53, 55, 57–58, 63–64, 66–68, 72–73, 75, 83–84, 89, 92–93, 97, 105, 108–110, 117, 127–128, 136–142, 145–147, 166, 185, 198–203, 209, 213, 226, 231–232, 234–235, 238–239, 244–245, 249
Yavapai County Chamber of Commerce 200, 227
Yavapai County Messenger 191, 221
Yavapai-Prescott Tribe 9
Yavapai Supply Company 53, 113, 119
Yavapai, the Land of Opportunity 136
Yolo Ranch 72
Young, Hiram 26, 32, 60
Young, Mrs. W.B. "Hattie" 73
Young's Farm 247–248
Yount, Dr. Clarence 38
Yuma 214, 230

Zurcher, Agnes 117

www.ingramcontent.com/pod-product-compliance
Ingram Content Group UK Ltd.
Pitfield, Milton Keynes, MK11 3LW, UK
UKHW050538150426
5217IPUK00026B/1983